ANALYSING ARCHITECTURE NOTEBOOKS

Architecture is such a rich and subtle field of human creativity that it is impossible to encapsulate it completely in a single book. I tried to describe some of the basics in an earlier book, *Analysing Architecture*, which has now appeared in four editions, increasing in size each time. But even though that book has almost doubled in content, there is more to cover. So, rather than make the original even heavier, I have decided to add further chapters as a series of separate smaller volumes.

These *Analysing Architecture Notebooks* are the new chapters I would have added to *Analysing Architecture* had not excessive size become a concern. The series format also allows me to explore topics at greater length than if I were confined to just a few extra pages in the original book. Nevertheless the shared aim remains the same: to explore and expose the workings of architecture in ways that might help those who face the challenges of doing it.

Simon Unwin is Emeritus Professor of Architecture at the University of Dundee in Scotland. Although retired, he continues to teach at the Welsh School of Architecture in Cardiff University, Wales, where he taught for many years. His books are used in schools of architecture around the world and have been translated into various languages.

Books by Simon Unwin
Analysing Architecture
An Architecture Notebook: Wall
Doorway
Exercises in Architecture – Learning to Think as an Architect
Twenty-Five Buildings Every Architect Should Understand
The Ten Most Influential Buildings in History: Architecture's Archetypes

ebooks (available from the iBooks Store)
Skara Brae
The Entrance Notebook
Villa Le Lac
The Time Notebook

The Analysing Architecture Notebook Series
Metaphor
Curve
Children as Place-Makers

Simon Unwin's website is at *simonunwin.com*
Some of Simon Unwin's personal notebooks, used in researching and preparing this and his other books, are available for free download from his website.

ANALYSING ARCHITECTURE NOTEBOOKS

CURVE

possibilities and problems with deviating from the straight in architecture

First published 2019
by Routledge
2 Park Square, Milton Park, Abingdon, Oxon OX14 4RN

and by Routledge
52 Vanderbilt Avenue, New York, NY 10017

Routledge is an imprint of the Taylor & Francis Group, an informa business

© 2019 Simon Unwin

The right of Simon Unwin to be identified as author of this work has been asserted by him in accordance with sections 77 and 78 of the Copyright, Designs and Patents Act 1988.

All rights reserved. No part of this book may be reprinted or reproduced or utilised in any form or by any electronic, mechanical, or other means, now known or hereafter invented, including photocopying and recording, or in any information storage or retrieval system, without permission in writing from the publishers.

Trademark notice: Product or corporate names may be trademarks or registered trademarks, and are used only for identification and explanation without intent to infringe.

Publisher's Note
This book has been prepared from camera-ready copy provided by the author.

British Library Cataloguing-in-Publication Data
A catalogue record for this book is available from the British Library

Library of Congress Cataloging-in-Publication Data
Names: Unwin, Simon, 1952- author.
Title: Curve : possibilities and problems with deviating from the straight in architecture / Simon Unwin.
Description: New York : Routledge, 2019. | Series: The analysing architecture notebook series | Includes bibliographical references and index.
Identifiers: LCCN 2018036919| ISBN 9781138045941 (hb : alk. paper) | ISBN 9781138045958 (pb : alk. paper) | ISBN 9781315171661 (ebook)
Subjects: LCSH: Architectural design. | Curvature.
Classification: LCC NA2750 .U584 2019 | DDC 729--dc23
LC record available at https://lccn.loc.gov/2018036919

ISBN: 978-1-138-04594-1 (hbk)
ISBN: 978-1-138-04595-8 (pbk)
ISBN: 978-1-315-17166-1 (ebk)

Typeset in Arial and Georgia

by Simon Unwin

for Nina

Devil's Bridge, Kromlau, Germany, nineteenth century

Curves reflect the sensible and the poetic in architecture. They may sometimes be characterised as subversive and decadent but they may also be seen as offering aesthetic value that transcends the pragmatic. The curve is both hero and villain, depending on your point of view.

CONTENTS

PREFACE 1
INTRODUCTION – INSTRUMENTS OF CURVATURE 3
ARCHITECTURE'S INNATE ORTHOGONALITY 7
THE EVER-PRESENT MELODY 25
MOVEMENT CURVES 41
STRUCTURAL CURVES 61
CURVES FROM STRAIGHT LINES 95
CURVES FROM NATURE 107
ORCHESTRATING CURVES 143
ENDNOTE 171
ACKNOWLEDGEMENTS 177
BIBLIOGRAPHY 178
INDEX 179

'Straight lines are easy... But as soon as you try to do a curve...'

Piers Taylor, in 'The House That 100k Built', BBC, 23 March 2017.

'Without it (the building of the rainbow bridge that should connect the prose in us with the passion) we are meaningless fragments, half monks half beasts, unconnected arches that have never joined into a man. With it love is born, and alights on the highest curve, glowing against the grey, sober against the fire. Happy the man who sees from either aspect the glory of these outspread wings.'

E.M. Forster – *Howards End* (1910), 2012.

'You may ascertain, by experiment, that all beautiful objects whatsoever are... terminated by delicately curved lines, except where the straight line is indispensable to their use or stability... Well, as curves are more beautiful than straight lines, it is necessary to a good composition that its continuities of object, mass, or colour should be, if possible, in curves, rather than straight lines or angular ones.'

John Ruskin – 'The Law of Curvature', in *The Elements of Drawing* (1857), 1971.

PREFACE

Baseball pitchers throw 'curve' balls, cricket bowlers bowl 'swing' balls… their intention is to bamboozle the batter (whom, in cricket, is advised to play with a 'straight' bat).

The metaphor 'straight' suggests rectitude and reliability. Incapable of being 'straight', curves are considered treacherous, they cannot be trusted; they tease and seduce to lead you astray. But in the twenty-first century, the curve has become something of a fetish for architects.

After the abiding orthogonality of Modernism, which dominated most of the twentieth century, curves have surged in popularity, enabled by advances in computer software and computer-aided manufacture and construction. But curves in architecture are not just a contemporary phenomenon; they have a long history, and come in many varieties.

There are curves generated by geometry. There are curves we draw with mechanical devices. Structures are curved to counter gravity. We move, we dance in curves. Ornament is curved. Curves embrace us. We enjoy being teased, confused, stimulated by curves. Nature displays curves that we want to emulate because we see them as beautiful or believe them to be right. We play with curves on computers, pulling and pushing them into sculptural shapes.

Like melody to rhythm, curves provide counterpoint to the straight and rectangular. Axial architecture makes us move in straight lines. 'Free plan' architecture provides a rectangular frame within which we are free to wander in curves. Curvaceous architecture holds the curve to itself. There are no right answers in architecture; there are adventures. And curves draw us into adventures.

This Notebook is divided into thematic sections. The aim is to explore the territory rather than commit it to consolidated theorisation. Examples are taken from across history and geography.

The majority of the illustrations were drawn by hand directly on a Wacom computer touch screen. Their first encounter with paper will be with the printing of this book. Since it is understandable that many of the drawings depict curves, I am particularly grateful to Adobe for introducing a 'smoothing' facility to the brush tools available in Photoshop. A few of the illustrations were generated and manipulated on Autodesk 3ds Max; I felt it was my duty, writing about curves, to experience something of the possibilities of working with parametric software. Other drawings were constructed using more primitive 'parametric' devices: bits of string, pins and pens; pendulous chains; random hand movements.

I began thinking it might be difficult to fill a Notebook with material about curves in architecture. Writing this Preface at the end of the process I find that some of the issues raised could fill Notebooks of their own. But nevertheless I hope the reader finds that what is included stimulates trajectories of creative reflection in their own minds. After all, these Notebooks are intended not merely to be receptacles of information but to provoke ideas that you can explore and develop for yourself.

William Blake, 'Los and Enitharmon', Plate 100 of *Jerusalem: The Emanation of the Giant Albion*, 1804–20.

INTRODUCTION
INSTRUMENTS OF CURVATURE

To create a straight line we use a ruler, a straight edge along which we draw a pencil or pen. But, both conceptually and practically, it is more difficult to draw a curve. If a ruler is the authority of straightness, what is the authority of a curve? Over the centuries, architects, engineers, builders… have used various devices for creating curves.

There is no such thing as a random curve. All curves, however complex they may be, are determined by the mechanics of their generation. Some curves are generated by natural processes – the geological folding of rock strata, the growth of shells, the petals and tendrils of plants, the swelling waves of the ocean… (See the drawings of Leonardo da Vinci on pages 26–7.) But purpose is generally an essential attribute of architecture. And so we architects have to resort to some method for the generation of our curves.

CURVE GENERATION METHODS
circles and squiggles

The beauty of a curve derives from the authority by which it has been generated.

 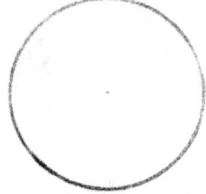 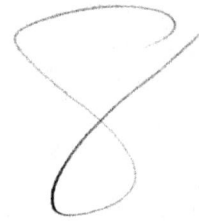

A hesitant curve lacks elegance. Its tentative ugliness derives from a lack of any controlling authority.

The beauty of a perfect circle derives from the consistency of its curve, which is a product of the mechanism by which it was drawn (below).

A rapid squiggle is usually more beautiful than a hesitant curve. Its authority derives from the confident gesture that drew it.

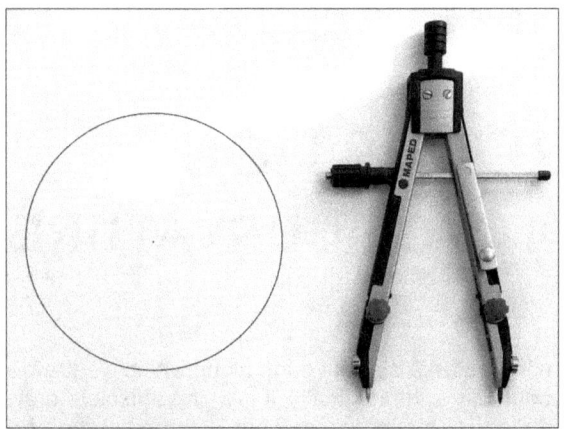

The curves of fluent handwriting (above) derive from the confidence of the hand that writes the word (even when the word itself is misspelt). As will be seen in the following pages, some architects have generated designs from gestural squiggles.

The pair of compasses controls the curve of the circle (above).

We can also draw circles with a piece of string, a pin and a pen (right). Large circles and other curved figures were, in ancient times, probably drawn on the landscape using pegs and ropes. Many of the examples of architectural curves in the following pages were made using compasses or ropes and pegs.

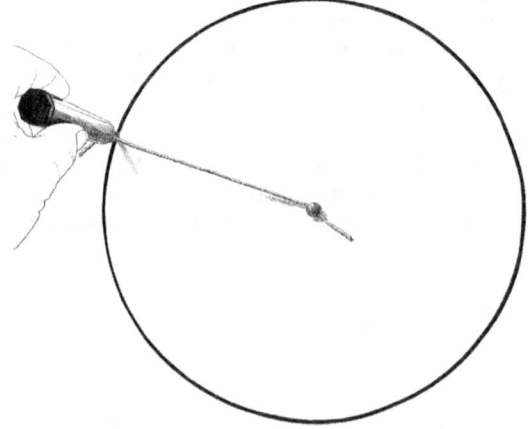

CURVE GENERATION METHODS
templates

We can also draw circles using templates such as a plate or saucer, or a specially made circle template (right).

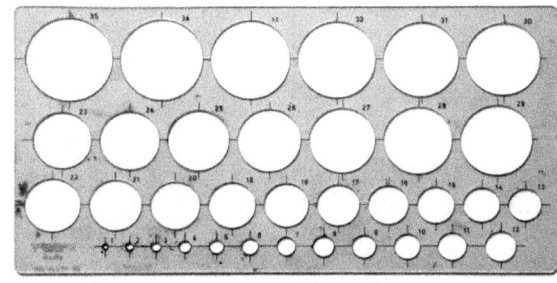

For more subtle curves, templates called French curves (right) have been used. Traditionally these were used by architects and engineers to produce smooth but non-regular curves, as in routing railway lines or designing classical ornament.

The tightening or relaxing lines of French curves are derived from the Euler spiral (below), the radius of which becomes progressively tighter.

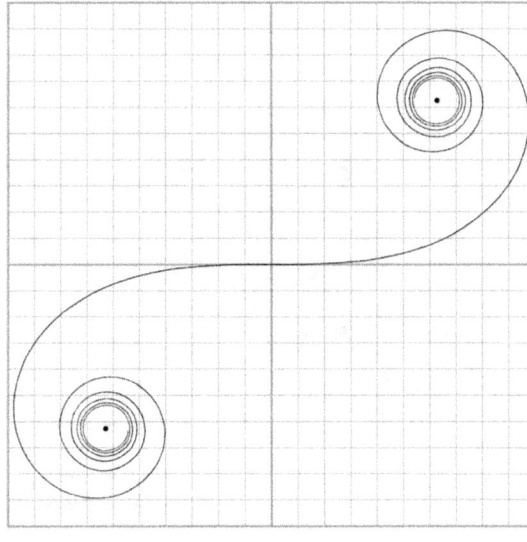

Architects have also used flexible bands or bars – flexicurves – that when bent into the desired curve remain fixed until bent again.

CURVE GENERATION METHODS
splines and parametrics

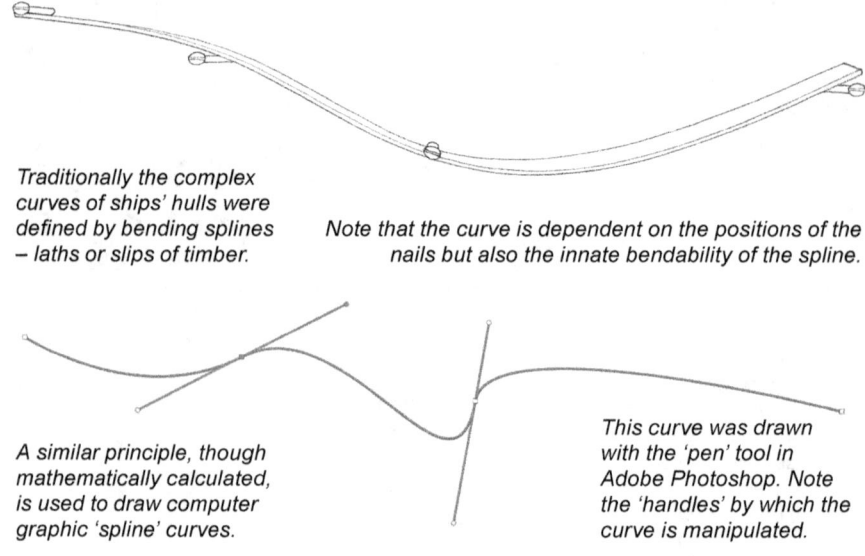

Traditionally the complex curves of ships' hulls were defined by bending splines – laths or slips of timber.

Note that the curve is dependent on the positions of the nails but also the innate bendability of the spline.

A similar principle, though mathematically calculated, is used to draw computer graphic 'spline' curves.

This curve was drawn with the 'pen' tool in Adobe Photoshop. Note the 'handles' by which the curve is manipulated.

These three pages have illustrated just some of the common ways in which architects generate two dimensional curves. You will find more in the following pages, and not only regarding two-dimensional curves... Mathematical 'spline' curves are also the basis for 'non-uniform rational B-spline' (NURBS) surfaces (see below and page 106). Manipulating these surfaces is one of the key techniques in computer parametric software.

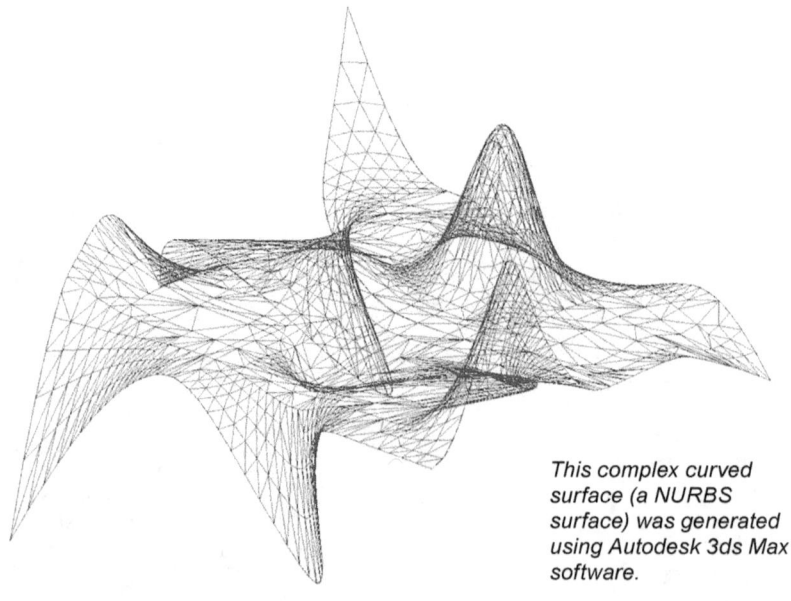

This complex curved surface (a NURBS surface) was generated using Autodesk 3ds Max software.

ARCHITECTURE'S INNATE ORTHOGONALITY

Many aspects of life are conditional upon right geometry. As TV detective Columbo (Peter Falk, above) discovered at the end of *How to Dial a Murder* (James Frawley, 1978), you cannot play pool with a curved cue, anymore than you could with a bumpy table or a cubic ball. This is only a small example, but generally things do not work properly if their geometry is wrong.

This Notebook explores the contributions curves make to architecture but there are lots of reasons why most of the world's architecture is orthogonal. Since ancient times most buildings have been built rectangular: their plans have right-angled corners; their walls are flat and vertical; their doorways and windows are rectangles; their corridors straight… Any consideration of curves has to be set in the context of the normative orthogonality of architecture. Here then are a some reasons why most architecture is NOT curvy.

SIX STRAIGHT DIRECTIONS
how our bodily presence makes the world orthogonal

Curves are, by definition, not straight. To understand the role and reputation of curves in architecture it is first necessary to understand some of the reasons why the straight line, right angle and flat surface have conditioned architecture through history.

It is a cliché to say there are no straight lines in nature. It is also wrong. Nature is replete with straight lines. Apart from the horizon across the ocean, slivers of sunshine striking into forest darkness through the leafy canopy, edges of crystal, the flatness of a salt lake... the most architecturally interesting are those we introduce into the world just by our own presence. We see in straight lines, stand up straight, and try to move in straight lines. These are all powerful influences on architecture.

The artist Antony Gormley has introduced casts of his own body into various natural and urban environments. Whatever other artistic merits they possess, these works are profoundly architectural. They stand for us all.

One of Gormley's most famous installations is called 'Another Place' (above; 1997). One hundred of his casts stand on Crosby beach in Lancashire. All stand upright and stare out to sea, facing in the direction of the setting sun. After twenty years they have suffered a variety of fates: the incoming tide submerges them by different degrees; shifting sands have buried them to different levels, even in some cases up to their necks, or elevated them above the sand's surface on their supporting piles; people have dressed them in shirts and bikinis; some have a acquired a rough crust of barnacles... But through all this – through all these indignities and apotheoses ('the thousand natural shocks that flesh is heir to') – each stands and survives as nothing less than a temple to human presence on the earth.

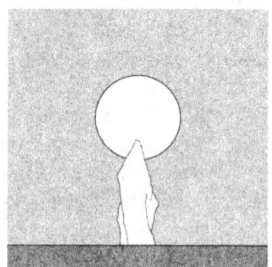

We are fascinated by alignments, and mark them with 'sights'.

A four-square relationship with the world is a defining characteristic of the human being.

Around the world, most temples have a simple underlying form: a rectangle with a doorway and housing an effigy or an altar (right). The geometry of this form resonates with our own; it is in harmony with how we project our own orthogonal form on the world.

We have eyes that look forward. We move forward. Forwards is our prime direction. Opposite this, behind us, is backwards. At right-angles are our sides – right and left. Above is above (usually the sky or heavens); and below is the ground (or the underworld). Our own geometry conditions how we see the world and how we shape (and make sense of) it with our architecture.

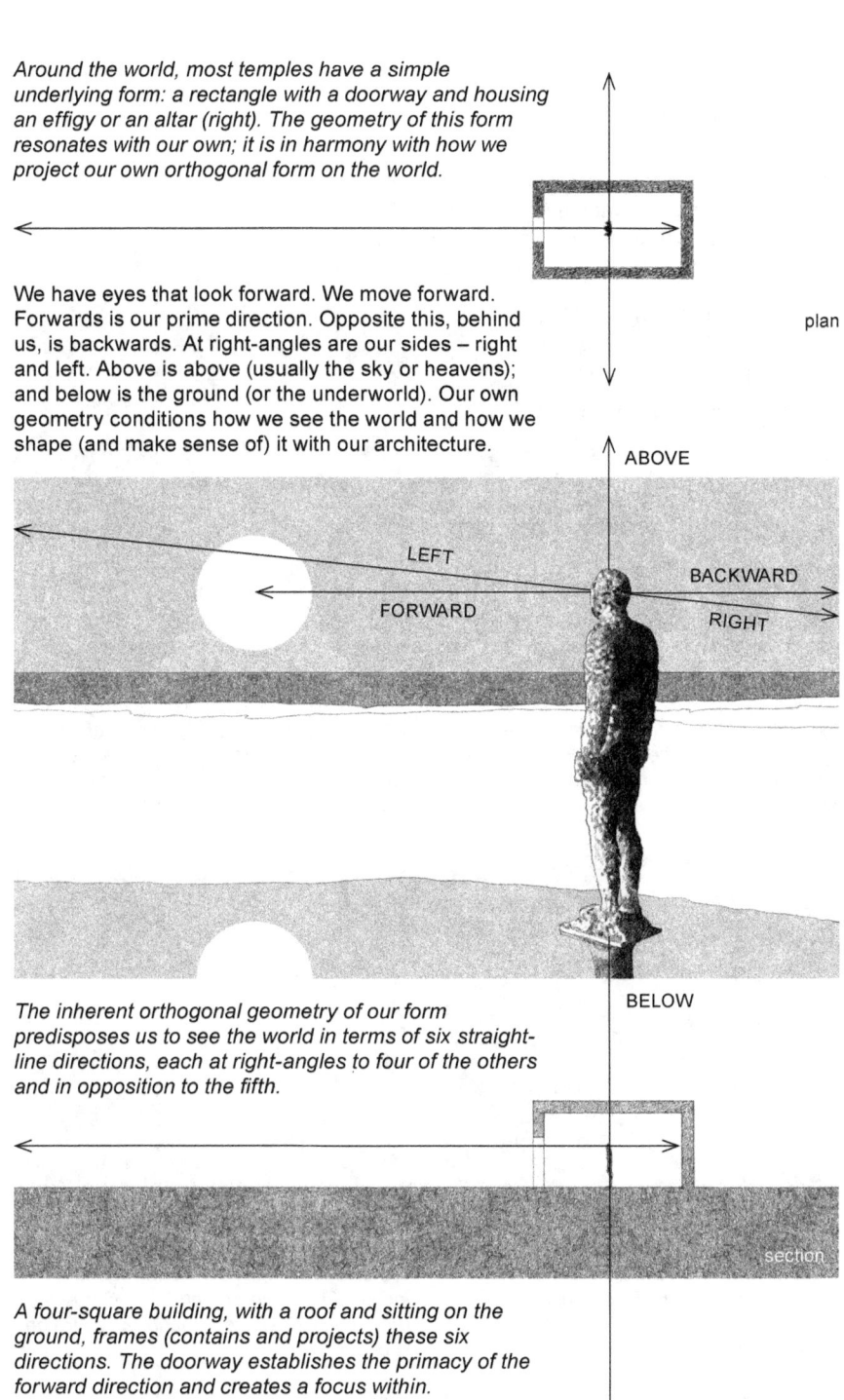

The inherent orthogonal geometry of our form predisposes us to see the world in terms of six straight-line directions, each at right-angles to four of the others and in opposition to the fifth.

A four-square building, with a roof and sitting on the ground, frames (contains and projects) these six directions. The doorway establishes the primacy of the forward direction and creates a focus within.

But own orthogonal geometry is not the only reason why architecture through the ages has been predisposed to the straight line, the right angle and flat surfaces.

CURVE

GRAVITY
why walls tend to be vertical

Our own innate geometry influences not only the ways in which we organise space in the world, it also conditions how we build structures. Just as we don't want to fall over, we don't want them to either.

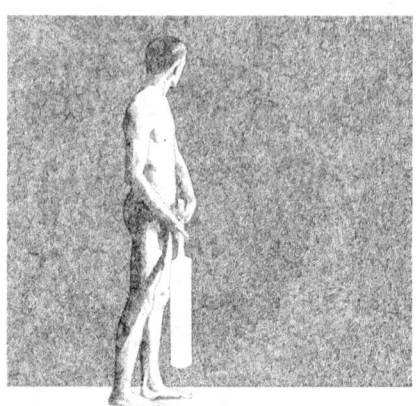

We empathise with verticality. When we stand up straight we know we are stable. Standing 'to attention' signifies we are strong and alert.

When we deviate from the vertical we know we are unstable. We are going to fall. We are not strong but vulnerable, and feel uncomfortable and unsafe.

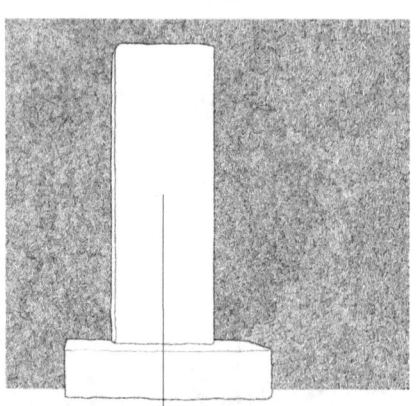

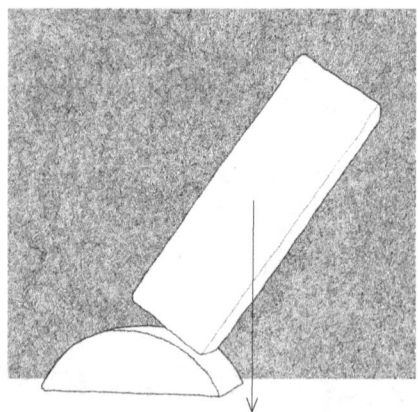

It is the same with buildings. We associate verticality with stability and strength. We call vertical lines 'true', implying that anything else is 'false'.

We also know that curves introduce instability and unpredictability. Curves unbalance; they are treacherous. Straight is best?

So the vast majority of buildings, right across the world, have perfectly vertical columns and walls. Their structural stability depends on it. Curves complicate things.

GEOMETRY – STONE
why materials tend to be rectangular

Over many centuries we have developed materials that make it easier to build square and vertical structures. Building with curved components in traditional ways is time-consuming and difficult.

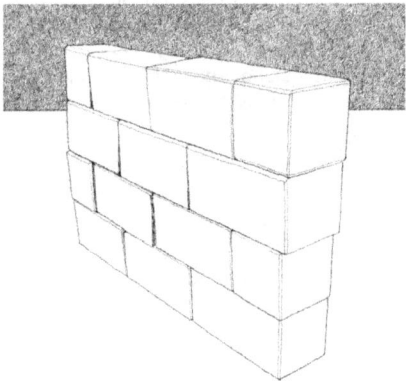

Even on a flat surface, it is almost impossible to build regular and stable walls from round pebbles. Without mortar, they are little better than heaps of stones.

Regular and stable walls are much easier to build with rectangular blocks. They can be built straighter, taller, and more quickly.

Walls built of round boulders tend to be thick and uneven, have to be carefully constructed, and have sometimes wide mortar joints.

Ashlar walls can be thinner and smoother, with straight narrow joints. In traditional architecture they are associated with more 'polite', higher status buildings.

The force of gravity is countered more efficiently by regular building components constructed into straight vertical walls with regular horizontal joints… and without curves.

GEOMETRY – BRICK
making bits fit together

'You say to a brick, "What do you want, Brick?"'   Louis Kahn, to students, 1971.

One answer the brick might give to Louis Kahn's question at the top of this page (but not the one Kahn himself suggested – see page 98), might be, 'I want to be built in straight lines in all three dimensions, and with right-angled corners. It's in my nature!...'

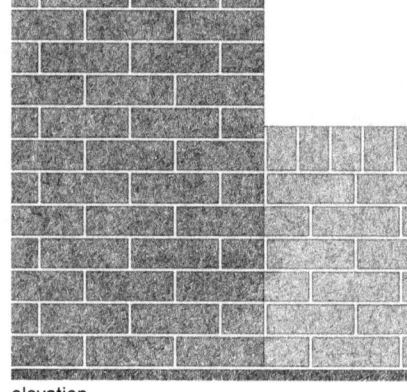
elevation

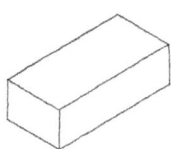

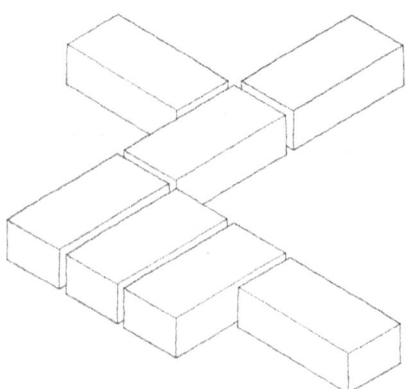

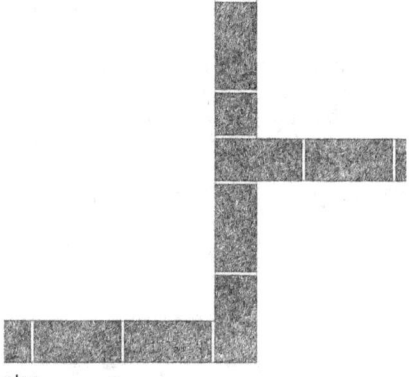
plan

'I myself am rectangular, with straight edges and parallel sides, and have flat faces, each at right angles to all but one of the others. Because of that I like to fit in neatly with all my identical friends. That way we naturally make straight lines, right-angled corners and vertical walls with flat faces. I don't really ever think about curves! They're not in my nature!'

Building walls is easier with standardised rectangular elements such as bricks. But they are predisposed to producing straight lines and rectangular openings.

GEOMETRY – TIMBER
straight and square makes building easier

Straightness and squareness constitute the normative principle of most human construction. Together they offer a reliable integrity that informs all neat and stable building. It is the same with timber. Wood might grow in curves, but in building many things are possible only with straight and square pieces of timber. Although straightness and squareness might seem to be rigid principles, they offer infinite variation in orthogonal form and easy replacement for repair.

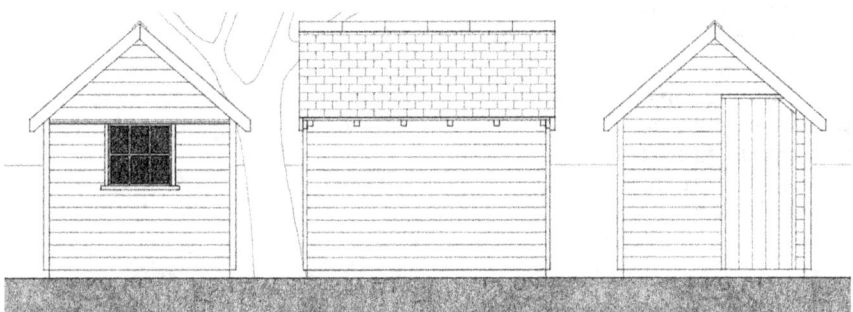

The building of a simple garden shed is only made simple by the availability of straight and square timber of different dimensions. Straight, square and parallel makes it easier to fabricate windows and doors that will open and close. Standardised straight and square components – whether studs, plates, braces, boards, rafters, slates or ridge tiles – are easy to replace. Curves would only complicate matters or make construction impossible.

Straightness and squareness can be components of elegant design. Care is taken to season timber before final planing and use, so that warping will not cause problems in construction or cause failure later.

GEOMETRY – CONCRETE AND STEEL
straight and square is neat and stable

Every page of this book is laid out on a framework of straightness and squareness. Not to do this would risk messiness and disorder. Such orthogonal neatness is a principle of the 'architecture' of this book. It is the same with the majority of built architecture. Right angles are neat and easy to construct.

The shape of reinforced concrete (above) is dependent on the formwork it is poured into. Since this is most readily constructed of sawn timber or sheet steel, it too, due to the geometry of making, is generally straight and square. The rectilinear geometry of the formwork conditions the geometry of the concrete.

Frame buildings, whether of steel or concrete structure, tend to be rectangular because of the straightness and squareness of their components.

Steel beams are manufactured straight because that makes them most useful; they are ready to be used in many different situations. Joined, straight beams produce rectangular structural frames (right), and hence straight and square buildings.

GEOMETRY – SPATIAL ORGANISATION
rectangular units do not waste space

Jigsaw pieces may have curved edges and fit together neatly to create a picture. But it is not so in architecture. The most efficient way to plan a composition of adjacent spaces is to make them rectangular. In that way, straight edges and right angles fit neatly together.

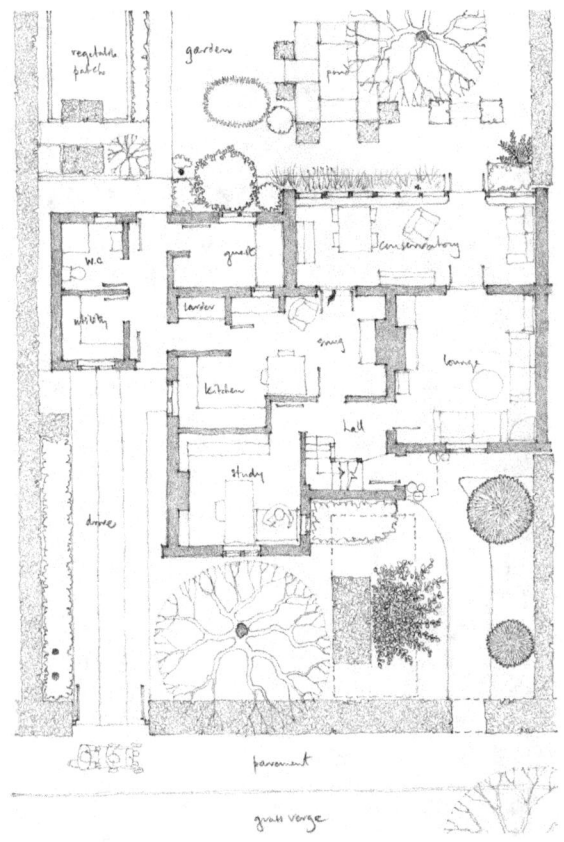

Like the vast majority of houses around the world, my house and garden (right) is a composition of rectangular spaces. Some are enclosed by walls; some overlap or open one into another; but they fit easily together because they are rectangular with straight sides.

Straightness and squareness make the accommodation of furniture easier too. Tables, cupboards, shelves... can all fit neatly against walls and next to each other.

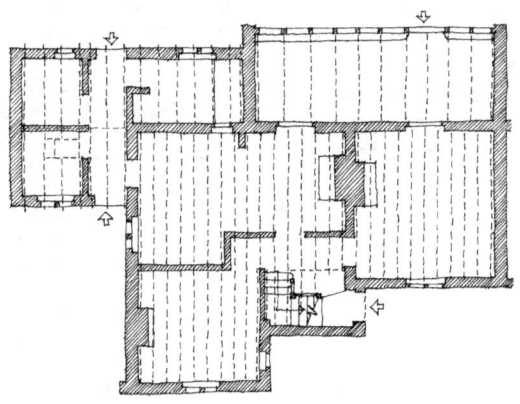

It is simple to put structure over rectangular spaces with parallel walls too. For the intermediate floors and the roof (right) straight timbers span with equal spacing across from one wall to another. All is straightforward.

GEOMETRY – FURNISHING ROOMS
rectangular things fit rectangular rooms

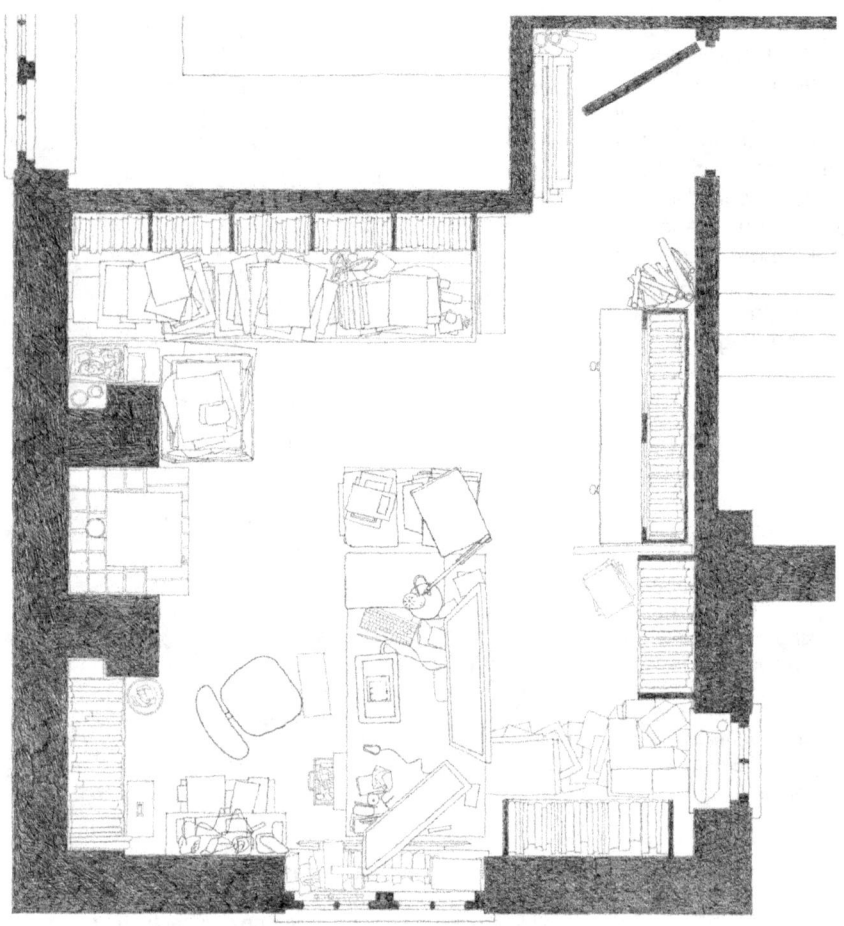

Because of the geometry of making and our own geometry, most furniture is made according to orthogonal principles. This also means it has a symbiotic relationship with a rectangular room.

Everywhere we live in rectangles. My study is even messier than this drawing suggests but, even so, the vast majority of stuff in it is rectangular. The room itself is roughly square; the windows are rectangular; the door would not work nor fit its doorway if it were not a rectangle. The hearth is rectangular. The stove, desk, drawers and bookcases are rectangular. The books themselves, framed pictures, and every sheet of paper in the room, are rectangles. That's why they fit neatly (!) together in piles and rows or hang flat against the walls. My keyboard, screens and scanner are rectangles, as are the boxes into which I throw random (and usually rectangular) clutter. There are a few circles – lamp, tubes, bin etc. – and my mouse, pens and swivel chair are composed of symmetrical curves, but only wires and wipe cloths introduce undesigned curves.

GEOMETRY – URBAN PLANNING
a grid city is a legible city

The neatness of orthogonal geometry – straightness and squareness – works at all scales (nearly). When cities are designed from scratch – rather than growing organically over many years – urban designers often choose to lay them out on a rectangular grid. Doing so means plot sizes are regular (and perhaps egalitarian), streets and avenues straight, and navigation simple.

Miletus (right) was a fifth-century BCE Greek city near the mouth of the river Meander now in Turkey. It is the birthplace of the urban grid, invented by Hippodamus, which brought geometric and perhaps also social sense to the layout of cities.

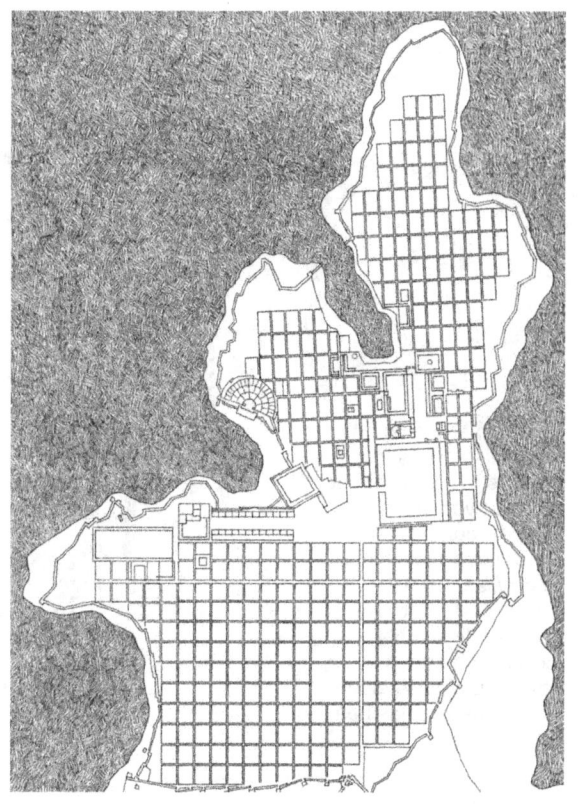

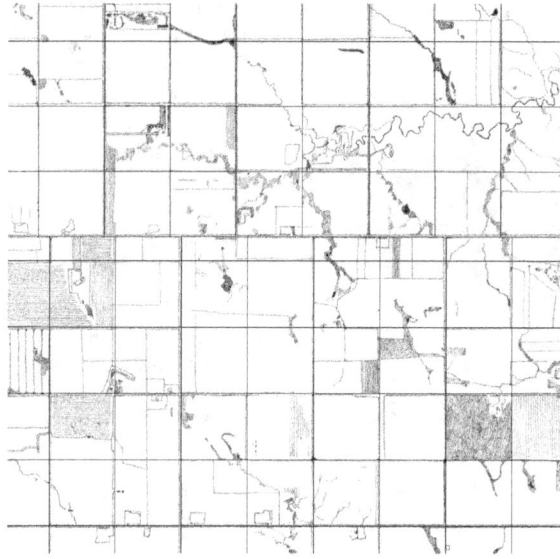

The straight line in the landscape is the quintessential sign of human presence. Since the eighteenth century, surveyors have laid out large areas of the plains of North America in rectangular grids. The grid allows roads to be straight for miles. The square parcels of land are more economical to plant, farm and harvest than irregular fields. (But the 'fault-line' visible in this image is caused by having to accommodate the grid to the curvature of the earth.)

AXIS OF SYMMETRY
the mirror line

Perhaps we empathise with symmetry because we are ourselves symmetrical.

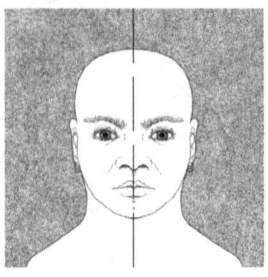

But symmetry offers its own satisfactions too. There is something neat and resolved about one side of a composition matching the other side exactly, as across a mirror. That mirror line is necessarily straight and has provided the axis on which much architecture through history has been designed.

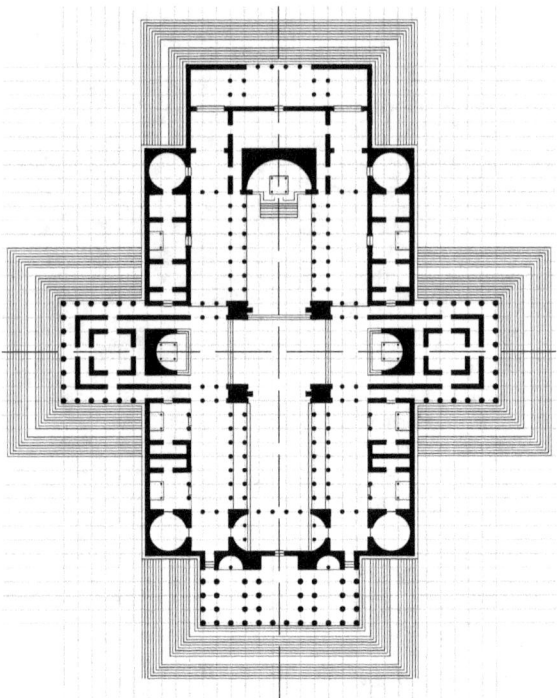

The axis line has dominated architectural composition since ancient times. Often generated by a doorway in alignment with a focus (an altar for example) the architectural axis resonates with our own forward direction. But it also provides a principle for composition.

When we stand on the axis of a building we feel linked to it. We see our own symmetry mirrored in its.

These effects depend on the power of the straight and intangible axis.

A CASE AGAINST CURVES
curves are decadent, subversive, morally suspect...

There is semantic resonance between geometric terms such as 'right', 'straight', 'square', 'level'... and moral rectitude and reliability; just as words such as 'bent', 'crooked', 'curved', 'warped'... evoke moral suspicion and distrust.

From the preceding pages, you can see that there are influences at play in the world (some might call them 'laws') that predispose architecture to orthogonality. Gravity, our own bodies, construction geometries, rational planning... all suggest that the sensible way to design architecture is with right angles and straight lines. These 'laws' constitute the case against curves in architecture. Sometimes it is presented as a moral case.

The Old Testament (below) suggests that straightness and squareness are more than pragmatically desirable characteristics; they should be counted as moral virtues.

'Prepare ye the way of the LORD, make straight in the desert a highway for our God. Every valley shall be exalted, and every mountain and hill shall be made low: and the crooked shall be made straight, and the rough places plain: And the glory of the LORD shall be revealed.' Isaiah, 40.

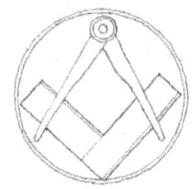

Presented in such terms, straightness and squareness are characterised as moral as well as geometric principles for establishing the ideal world. The symbol of Freemasonry, for example, includes a set square, as well as a pair of compasses (right). Utopia (paradise even), it is suggested, is flat and orthogonal, with straight roads and prismatic cities. According to this ideology, architects should aspire to be true to their moral vocation as agents of right angles and straight lines. Many have.

Architecture that is rectilinear has been presented as rational and pure... – the antithesis of curvy architecture, condemned as subversive and decadent.

Robert Mallet-Stevens, Villa Noailles, Hyères, 1923

CHALLENGING THE ORTHODOX
Twisted Farmhouse, Pennsylvania, Tom Givone, 2017

Online architecture news websites (such as Dezeen or Archdaily) imply that most contemporary architecture (in 2018) is curvaceous. There is a general fascination with curves. Architects play with parametric computer software; builders charge large sums of money to realise the designs generated; and clients, paying those large sums, welcome the publicity their curvy buildings receive.

The Twisted Farmhouse, by Tom Givone, is a reworking of a traditional 1850s timber frame house in rural Pennsylvania. Following the 'gravitational pull' of the orthogonal geometry of making, the original building is four-square. But the extension is curvy. At least it appears so; inspection of the plans shows that all the rooms are, for practicality's sake, rectangular. It even seems that the extension, which appears curvy in profile, is in plan composed of layers of straight lines meeting in a corner that, as it rises, gradually deviates from the right angle.

Orthogonality prevails in the Twisted Farmhouse's kitchen (below) where there is a more conventional relationship between curved and straight. Here rectangular windows frame irregular nature outside. Curved vases, a loaf of bread, the grain of the timber floorboards and the veins in the marble sink, all offset the general orthogonality of the room and its components. The rooms and furniture work together harmoniously because they conform to the practical convention of rectangularity.

plan

kitchen

AN ORTHOGONAL CORE
Guggenheim Museum, Bilbao, Frank Gehry, 1997

Even one of the most celebrated of curved buildings – Frank Gehry's 1997 Guggenheim Museum in Bilbao – has a core of rectangular rooms. Whereas sculptures might sit well in curvaceous spaces, paintings hang better on flat vertical walls. And the building's curvaceous titanium panels are supported on a complex structure composed of straight steel beams.

Though the fame of the Guggenheim Museum derives from its curvaceous external form – which contrasts dramatically with the general orthogonality of the city around – and from the sinuous irregularity of some of its galleries (right), Gehry's distorted geometry is anchored to a core suite of rectangular rooms with flat vertical walls. Even its curvy walls are supported by hidden structures composed of straight pieces of steel.

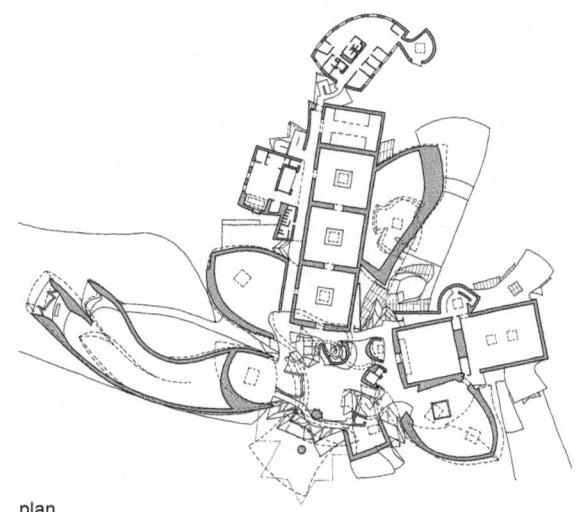

plan

DISTORTION
left-handed Rietveld chair, Julien Berthier, 2007

Transparent materials such as glass and water can distort straight lines into curves. The effect makes us wonder whether we ever see the world right; whether the 'straight' and 'true' is always distorted by the filters of our own perceptions

Old glass often has imperfect surfaces that distort the passage of light. This can mean that, through a multi-paned window, you see a slightly different world in each pane. Straight lines are displaced and distorted into curves. There is no reality through such glass. Each pane offers a subtly different distortion.

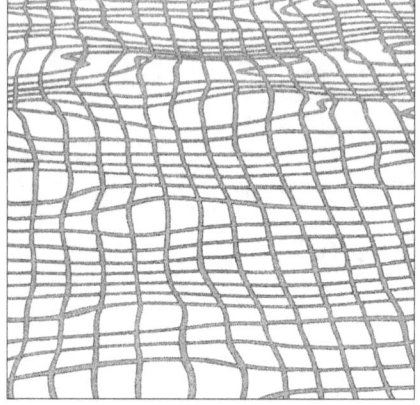

A uniform grid of square tiles becomes interesting by being distorted by water. Straight lines become curves. There are no longer any right angles. By distortion, standard elements become individual, each different from all others. It would take great effort to recreate this pattern in tiles without the help of the water.

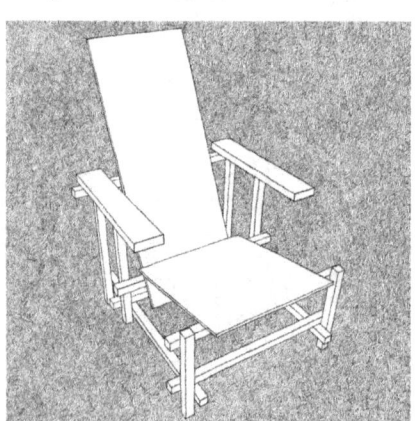

Gerrit Rietveld built his Red Blue Chair in 1918, just after the First World War. He wanted to make a chair out of straight rectangular pieces of timber, without any artful and curvy ornament.

In 2007, right-handed Julien Berthier drew the Red Blue Chair with his left hand. In doing so he distorted the straight lines into curves. It was much more difficult to construct.

JOKE CURVES
Krzywy Domek, Poland, Szotynscy Zaleski, 2004

Jokes work by throwing in a 'curve' at the end of a short narrative. The punchline adds a twist that surprises and amuses us. Curves can work like that in architecture too. Above is a building in Sopot, Poland – the Crooked House. It is distorted as if seen through an irregular block of glass or in a fairground mirror. Because of this it is one of the most photographed buildings in Poland.

Below, I have tried to draw a conventional version of a similar building. It is unlikely that many tourists would want to photograph it. Without the curves the architecture of the building is devoid of a punchline.

CURVES AND EXPENSE
for beauty or expression of status

There are factors that lead us to conclude, generally, that the sensible way to do things is in straight lines and right angles. And if we look at buildings around the world, this, in the main, is what we do. But we do not want to be sensible all the time; sometimes we want to transcend the sensible. Rather than walk with a steady regular pace, we sometimes want to dance… even if it might seem foolish.

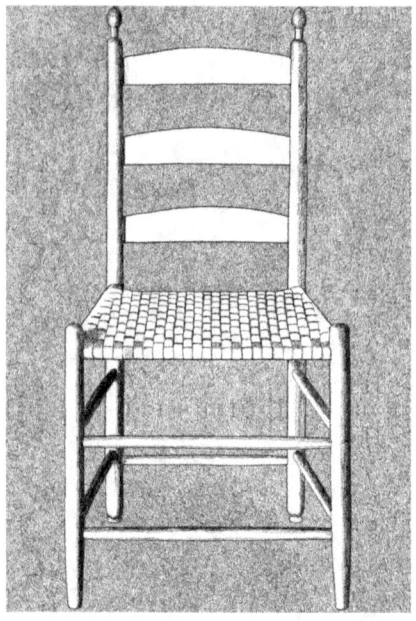

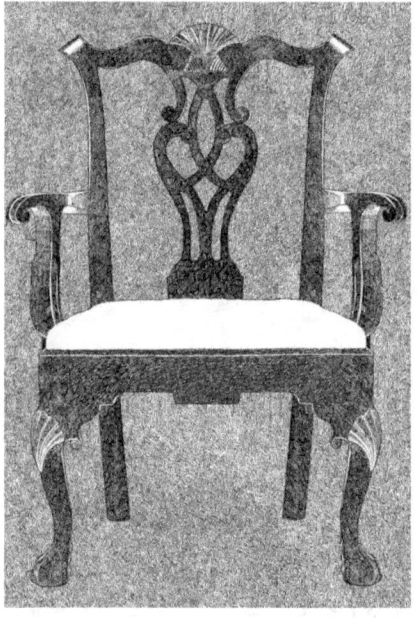

Relations between 'curved' and 'straight' are complex. 'Straight' – as in the Shaker chair above – can be simple and elegant, and may be interpreted as an expression of reserved humility and honesty.

Whereas being able to afford a curvy Chippendale chair (above) might be interpreted as a sign of wealth and status (independent of whether it is more beautiful than a simple chair).

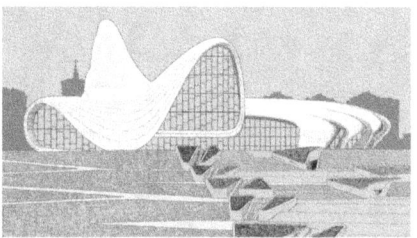

It is the same in architecture. Even a grand building like Mies van der Rohe's Nationalgalerie in Berlin (1968) can be a manifestation of honesty and austerity because of its strict adherence to regular geometry, the straight line, and rectangularity.

Whereas a desire to assert sophistication flamboyantly can lead to architecture which is ostentatiously expensive due to the complexity of the construction of its dramatic curves. This is the Heydar Aliyev Cultural Centre in Baku, Azerbaijan, by Zaha Hadid (2012).

THE EVER-PRESENT MELODY

It is easy to focus on the straight and square, and neglect the fact that architecture can never escape the curve. Architecture is delimited by curves, frames curves, is infused by curves, set in juxtaposition to curves, accommodates curves. If you accept the decadent nature of curves, then architecture is irredeemably infected by curves. Alternatively, you can see curves as the leavening of orthogonal architecture, the melodies that lighten a rigid beat. It is not as if the world, the universe, is short of curves. Even when architecture is scrupulously orthogonal, curves are its constant companion and foil. The world around is replete with curves: curves in topography; curves in flora and fauna; movement curves... The universe is full of curves: orbital curves; gravitational curves; even space itself, so we are told, is curved; as is the building block of all life, DNA. Architecture's main content – human beings and our activities, possessions, stories... – is a complex of physical, dynamic, narrative curves: our limbs move in curves; our organs are curved; we move in curves through space; our hair is curly. Architecture occupies a conceptual middle ground – it is always an interface – interposed between the vast and the intimate dominions of the curve.

ASSIMILATING THE CURVES OF NATURE
Leonardo da Vinci, observer

Out of curiosity and to make himself a better painter Leonardo da Vinci studied curves in many aspects of nature: in human form, curly hair, animals, plants, the folds of fabric, the movement of water... He drew on these studies in his own inventions.

HIDDEN GEOMETRY AND ORTHOGONAL FRAMING
Leonardo da Vinci, analyst and architect

But Leonardo also mused on whether the curves of nature were governed by an unseen armature of orthogonal geometry. He analysed the human head and body to see if it could be reduced to squares and other geometric proportions.

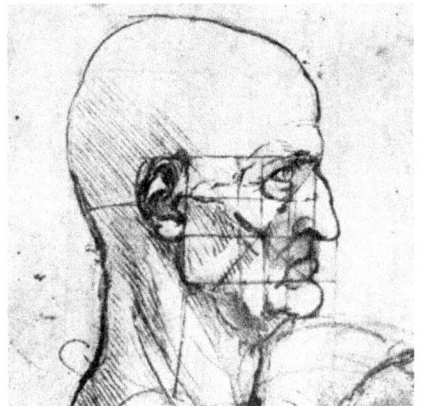
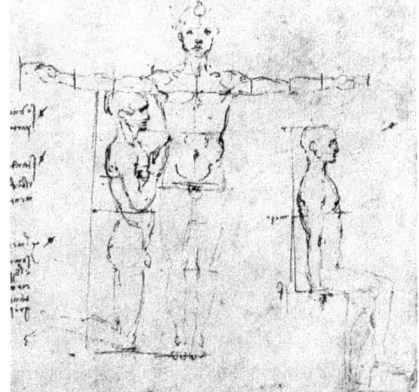

Leonardo saw 'right' geometry – whether a picture frame, a plinth for a sculpture, or enveloping architectural structure – as a foil for human and animal form, for nature, and for movement in space. In these ways the human mind invested the natural world with its own assumed touchstone – orthogonal geometry.

CONTEXT CURVES
architecture's setting is curvy

However regular a work of architecture, its regularity will always, at some point, hit up against the irregularity of the world.

At Miletus (page 17 and right), Hippodamus's grid is delimited by the city wall that follows the line of the coast at the time. (The Meander is now silted up, so the coastline has moved.) At its margins, the city is not regular, but consists of an interplay of grid and irregular border.

Below is part of Giambattista Nolli's 1748 plan of Rome. The line of the mid-twentieth-century Via della Conciliazione is dotted in. You can see that, even with its later extension, the axis that emerges from St Peter's Basilica soon hits the curving River Tiber.

Some might say architecture benefits from such interplays between regularity and irregularity.

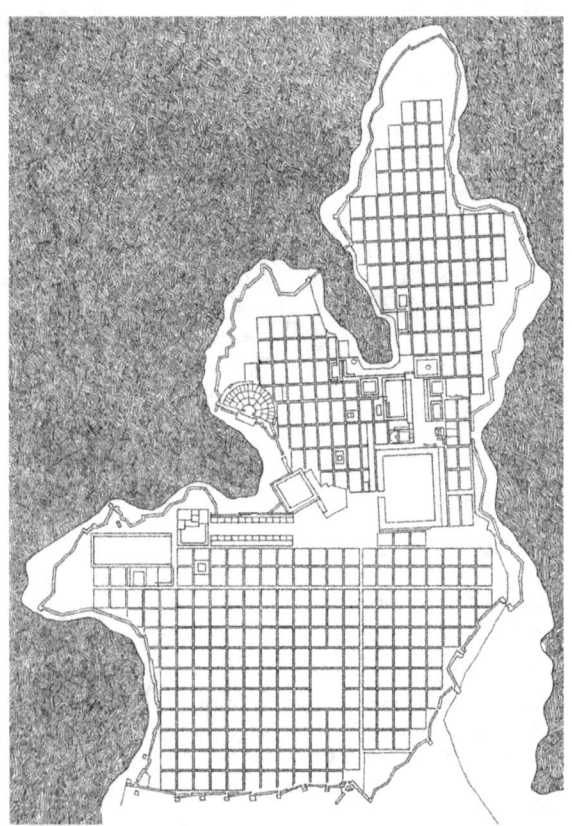

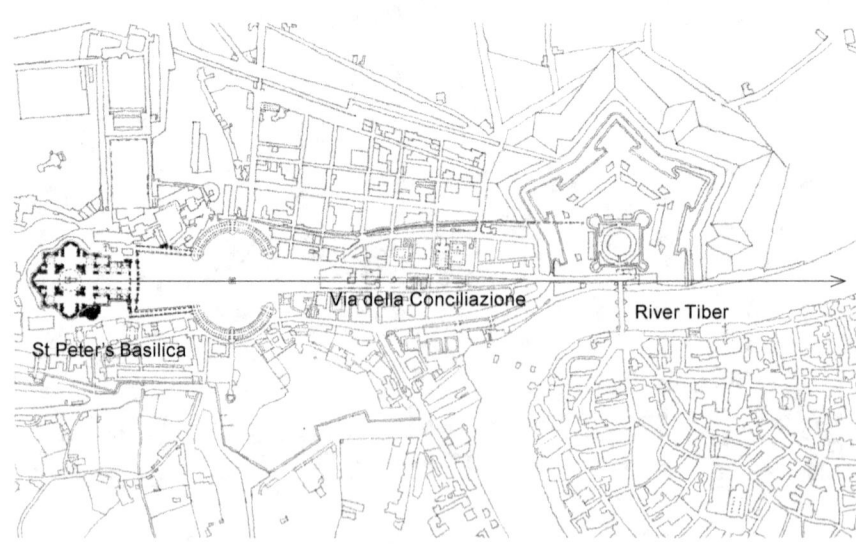

CONTEXT
complementarity

A temple is not complete without its context. Its orthogonal form derives its power from juxtaposition with the landscape. The full work of architecture comprises the interplay between the building and the surroundings. The axis of the temple projects as far as the horizon and beyond, linking the building and the presence of the god to the remote. But the prismatic geometry of the temple also contrasts with and complements the hills, valleys, rocks and trees around. You reduce the architecture of a temple by divorcing it from its surroundings, either by removing it to a museum or (as has happened in the case of the Bassae temple above) encasing it in a protective tent (however necessary).

The regular orthogonal (but now ruined) form of the fifth-century BCE Doric temple at Bassae in Greece (above) stood in contrast with the irregular landscape around. Symbolically, the perfect temple represented the presence of the god (and the mind of the architect) in the natural world. Its power derives from the juxtaposition of regular geometry with irregular nature.

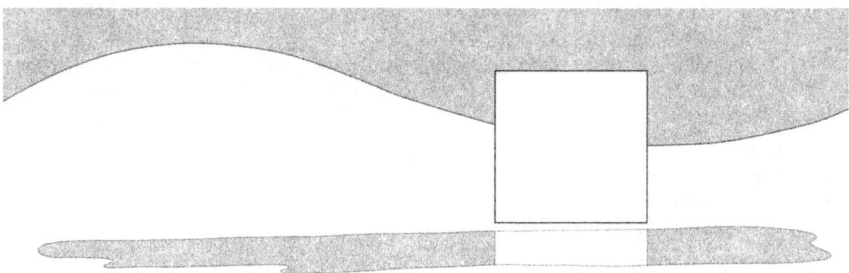

Platonic geometry and irregular 'curvy' nature are mutual foils. Their aesthetic and symbolic power can lie in their complementary contrast.

CONTENT CURVES
architecture accommodates curves

Curves intrude into the most regular and orthogonal of architectural constructions.
 Looking again at some of the examples given earlier in this Notebook (pages 13, 14 and 16):
 – the attraction of board-marked concrete (right) derives in part from the traces the grain of the timber formwork leaves on the surface of the concrete. The subtle curves of the grain markings offset the regularity of the rectangular timbers and the general flatness of the concrete.
 – the elegance of traditional Japanese timber construction (right) relies on its precision and scrupulous rectangularity. But it is also enhanced by the curving grain of the sawn, planed and smoothed timber as well as the occasional shake or crack. The poetry lies in neither alone, but in the juxtaposition of regularity with inherent curves. (Imagine exactly the same composition of elements constructed in perfect white melamine faced board. Something would be lacking.)

THE DANCE OF LIFE
our movements are curved

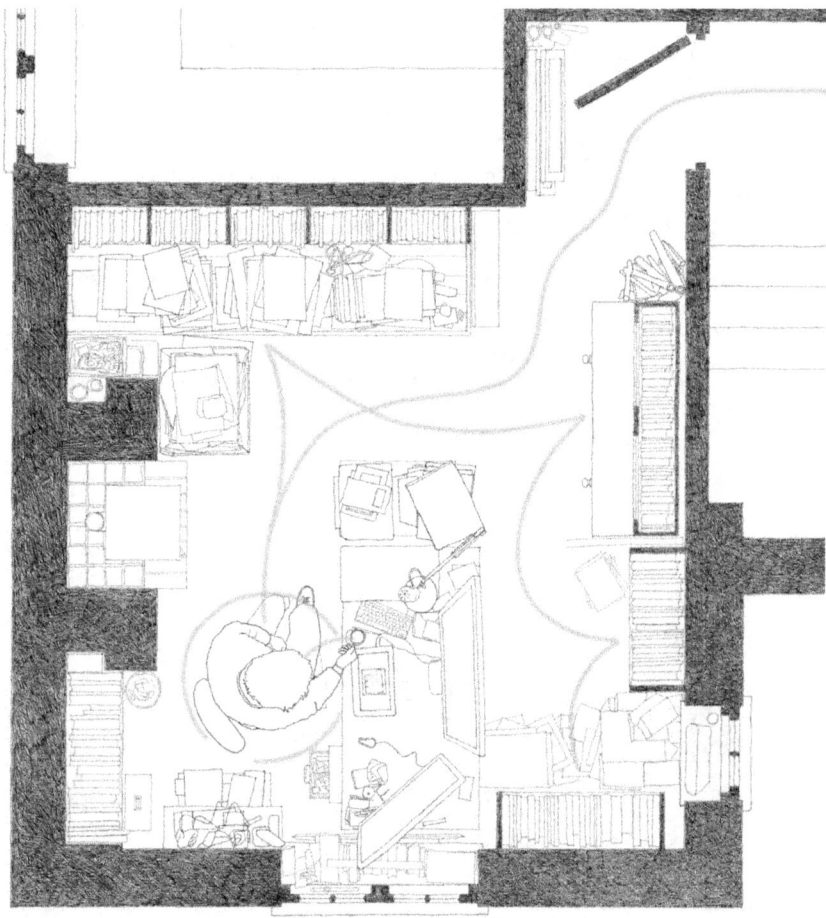

And in the rectangular world of my messy study, my movements to access books and papers traces choreographic curves. My being, in this mess of rectangles, is curved. My route is curved. My feet swing in curves as I walk. My chair swivels in curves. When I lift my mug of coffee to sip, it traces an arc from desk to mouth. We may think that the defining mark of human presence in the landscape is a straight line, or a prismatic form, but it is in the nature of our being and moving to be curvy. The rectangles of architecture are interposed between the curvaceous universe outside and our own curvy lives inside. Which poses a question: how curvy can architecture itself afford to be?

The spatial organisation of architecture sets rules (constraints) for our freedom of movement. In my study, my scope for movement is constrained by the modest size of the room and the excessive amount of furniture and clutter it contains.

FUSION
orthogonal geometry and curves combined

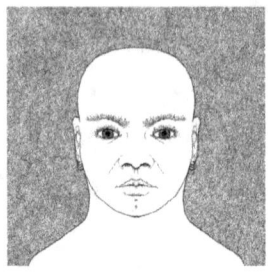

Our own faces are a fusion of curves and axial geometry. Their composition may be ordered by other kinds of geometry too.

Orthogonal geometry may frame curves. Curves may provide a complementary setting for orthogonal geometry. But orthogonal geometry and curves can also fuse.

Our own form is usually counted a prime example of the fusion of curves and regular geometry. Though we see living bodies as curvaceous, adopting asymmetrical poses and moving constantly, Renaissance theorists (following the Roman architect Vitruvius) argued that their underlying form was ordered according to the geometry of squares, circles, axes and Golden Sections. Consequently, it was suggested, the beauty of the perfect human form derived from a fusion of the two. The assertion of this principle of beauty influenced Renaissance architectural form.

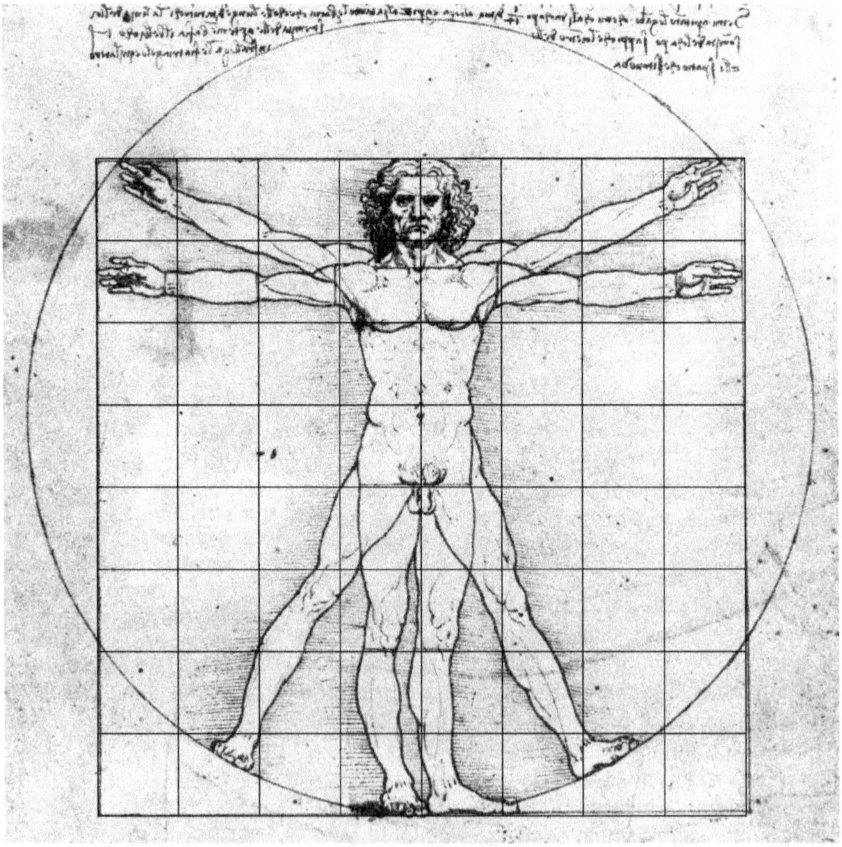

The power of Leonardo da Vinci's drawing of Vitruvian Man lies in its inference that human form is a fusion of curves and Platonic geometry.

FOIL
curves interacting with orthogonal geometry

All creative disciplines play, for aesthetic and poetic effect, with the juxtaposition of curves with regular geometry and axiality. In poetry, a classical 'geometric' form is the iambic pentameter – five sets of alternating unstressed and stressed syllables:

'The CURfew TOLLS the KNELL of PARTing DAY'
The first line of Thomas Gray's 'Elegy Written in a Country Churchyard', 1751.

It is against this geometric beat (like that of the heart) that the 'curving' poetic narrative is set. It is the same in music: curving melodies are set against the rhythmic geometry of a steady beat.

One of the classic strategies in hairstyling is to offset the axial geometry of the face with asymmetric curves.

'Jason', Simon Unwin, 1971

The same device is employed in the visual arts too. The curving lines of an abstract expressionist painting (left) are contained by the rectangular edge of the canvas and the frame. Its four-beat regular 'rhythm' contains the 'melody' of the curving paint strokes. The frame establishes a constraining horizon and a foil against which the energetic irregular composition is set. And in containing the curves it gives them their own world, separated from everywhere else... like a dancer dancing in an enclosed rectangular room.

The tokonoma (right) is a special place in a traditional Japanese house. A shrine to aesthetics, it is a niche in the orthogonal geometry of the house, open to but separate from the main living space. Within are displayed special objects: maybe a bonsai tree, a fine piece of pottery, a flower arrangement, a calligraphic scroll, some pebbles arranged in a tray of sand like a meditation rock garden... The curves of these objects stand, like in painting, as a foil to the rectangular discipline of the niche and the house.

CURVES AND THE APPEARANCE OF MOVEMENT
static and dynamic forms

Phrasikleia Kore, Ariston of Paros, 540 BCE

'Berlin Dancer', 2nd C CE

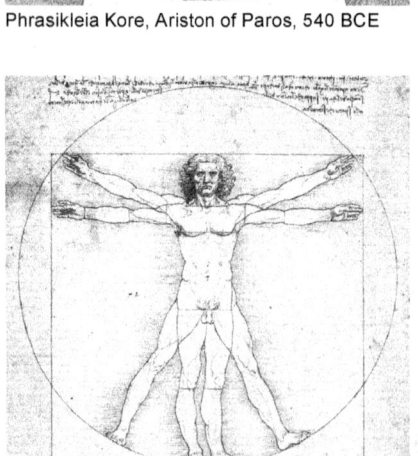

Leonardo da Vinci's 'Vitruvian Man'

The composure of the lady above left depends on her upright stance. Gentle curves contribute to the beauty of the sculpture, to the lady's femininity and her regal bearing, and hint at the flexibility of the fabric of her dress. But, even so, her demeanour is established by her perfectly vertical stance. She stands firmly and still on the ground. In contrast, the dancer above is light on her feet and caught in movement. Her elegance and vitality is instilled in the sculpture mainly by the expressive curves of limbs and the breezy fabric of her dress.

Vitruvius's description of human form is geometrically determined and static.

The straight upright line of a guardsman standing to attention expresses steadfast passive aggression while the dynamic curving lines of a ballet dancer express changing emotions. Upright straightness is sometimes cast as a 'male' attribute and curvaceousness as 'female'. But the images on these two pages suggest the dichotomy is not that simple; the characteristics and associations of straight lines and curves are more complex. Straight lines are static and dependably consistent. Curves are associated with movement and change.

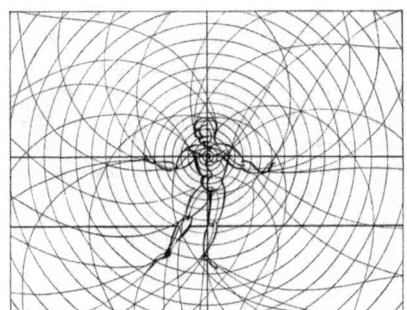

Oskar Schlemmer, from *Mensch und Kunstfigur*, 1924

Oskar Schlemmer saw human presence as defined by its curving movements in space.

CURVE

STRAIGHT LINES AND CURVES IN ARCHITECTURE
two contrasting towers

Pirelli Tower, Milan, Gio Ponti, 1958

Raffles City, Hangzhou, China, UNStudio, 2017

As with the images on the two previous pages, the elegance and stillness of Gio Ponti's Pirelli Tower in Milan (above) contrasts with the sense of movement instilled in UNStudio's design for Raffles City in Hangzhou (above right). The towers of the latter seem to sway and gyrate in relation to each other like dancers.

There is a great deal more to curves in architecture than sculptured curvaceous form suggesting movement. Architecture accommodates actual movement, which introduces factors that cannot be seen but are experienced. This theme will be explored in the next chapter. But before we move on, the following few pages illustrate some relationships between the curved and the orthogonal, looking in particular at the role of architecture as mediator between outside and inside.

RECTANGULAR FRAME AS INTERMEDIARY
Schröder House, Gerrit Rietveld, 1924

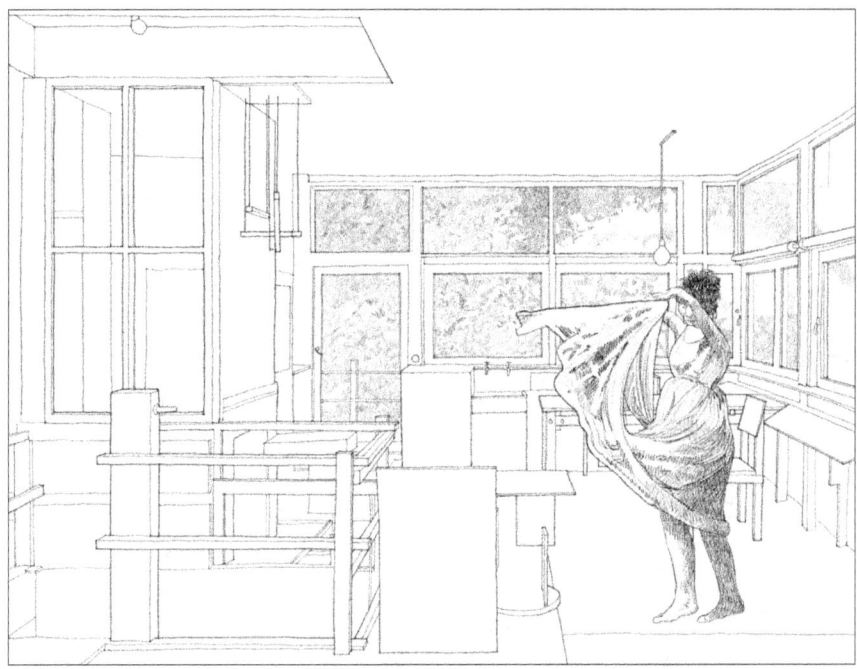

The Schröder House is a composition of rectangles arranged in three orthogonal dimensions to contain domestic life. In itself it is purely rectangular. But this composition is only completed by its essential ingredient – the person – who introduces the curves of human form and movement.

At the Schröder House there is a particular tree (right), whose intricate curves provide a foil to the rectilinearity of the house – like a bonsai in a Japanese tokonoma (see page 33).

Curves and rectangles juxtaposed act as symbiotic foils. Each offsets the other, relieving monotony, generating visual poetry. Often trees, and always people, provide the complementing foil for rectilinear architecture, which acts as intermediary between the two.

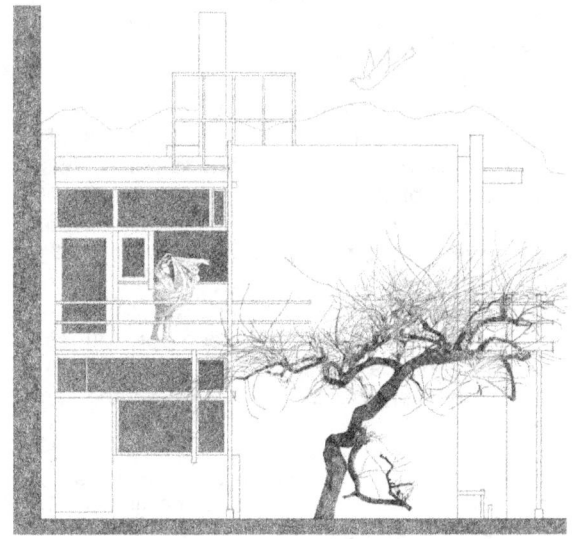

CURVE

BRINGING OUTSIDE CURVES INSIDE
Norwegian Wild Reindeer Pavilion, Snøhetta, 2011

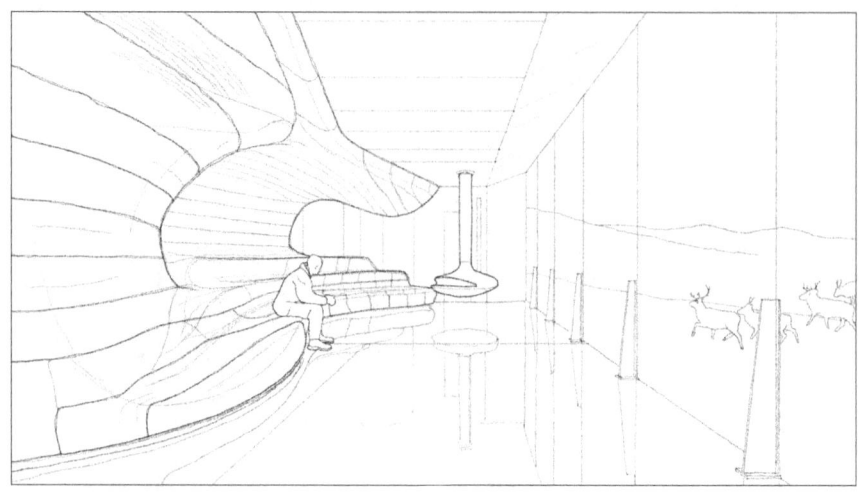

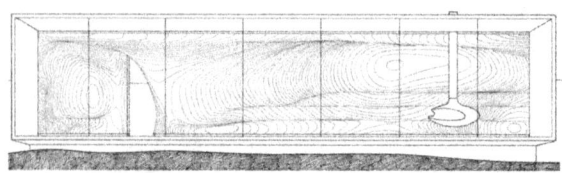

elevation

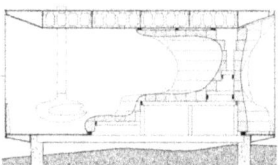

construction section

plan

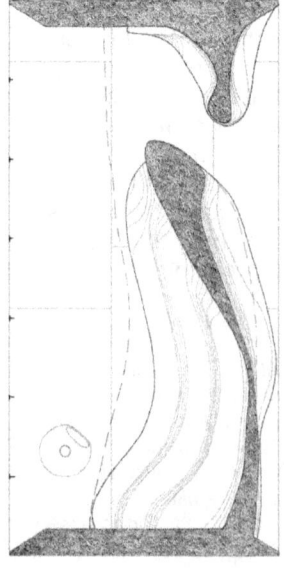

In their design for a viewing pavilion at the Norwegian Wild Reindeer Centre at Hjerkinn in the Dovre mountains, architects Snøhetta have inverted the usual relationship between architecture and the land.

Designing a refuge from which to watch reindeer, Snøhetta have constructed the form of a cold snow cave from baulks of warm timber. This curvaceous construction, incorporating seats and echoing the form of a refuge in the snow and ice, is contained within the strict orthogonal geometry of a glass fronted box.

A piece of nature – offering welcome refuge in the wilderness – has been transformed from snow and ice into timber, and framed in the quintessential architectural form of a rectangular box.

(Notice, in the construction section above right, that the timbers are shaped from rectangular baulks and supported on a rectangular framework of supports.)

WRAPPING CURVES OUTSIDE
Hotel Marqués de Riscal, Spain, Frank Gehry, 2006

By contrast (with the Norwegian Wild Reindeer Centre opposite), the American architect Frank Gehry has produced a number of designs in which curvaceous forms are applied as a wrapping to the outside of buildings.

Gehry's Hotel Marqués de Riscal in Elciego, for example, has a complex composition of rectangular accommodation blocks around which are wrapped huge multi-coloured 'ribbons' of curved metal sheeting that, though fixed, appear to be billowing in the breeze.

These ribbons do not accommodate human presence in the way that the curves of Snøhetta's reindeer watching pavilion do. They provide some shade in the hot Spanish summer sun, but their main purpose is to give the building a distinctive and photogenic appearance.

The Hotel Marqués de Riscal is wrapped in curved metal 'ribbons', apparently frozen in a moment whilst billowing in the wind. The ribbons dress the building in a manner reminiscent of fashion design. By attracting attention they advertise the enterprise of both hotel and architect.

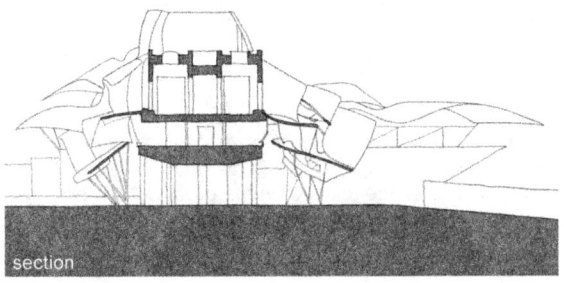

At the core of the building is a complex composition of rectangular blocks that house the hotel's accommodation and reception areas.

CURVING THE SPACE OF LIFE
Okurayama Apartments, Japan, Kasuyo Sejima, 2008

The usual dwelling contains orthogonal rooms and is set in irregular surroundings. Sejima's design for apartments in a suburb of Yokohama sets curvy rooms and courtyards within an almost rectangular city block.

We are used to living in orthogonal spaces. Our own inherent orthogonality – our six directions and centre – seems to harmonise with those of six-sided rooms. Furniture tends to fit straight walls and rectangular spaces too. But Sejima turns these considerations inside out. The life of the residents of these apartments has to find its order in curved spaces.

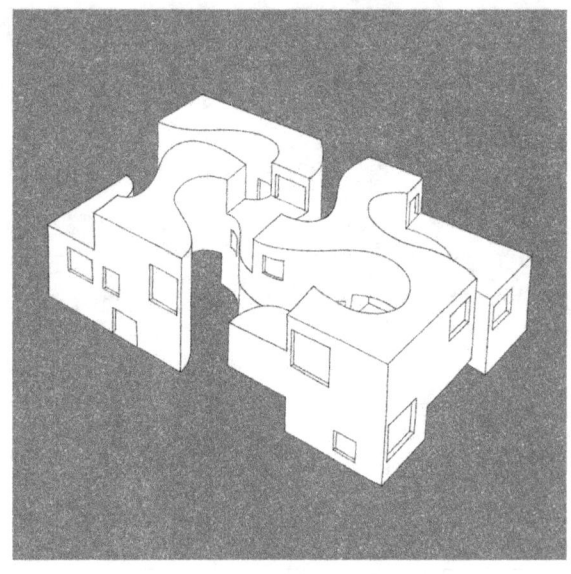

The apartments are closely surrounded by more orthodox city blocks (not shown). The curvy inlets of space, excavated from the building's three-dimensional volume, mean that no apartment is anywhere more than a few metres wide. Light and air are always accessible.

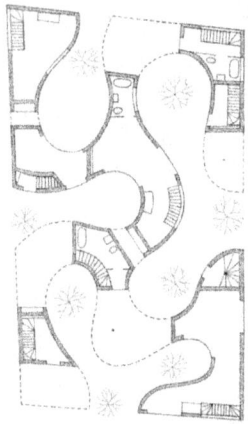  

ground floor   first floor   second floor

There are nine apartments, which twine under and over each other. The three floors have the same curvy footprint but the distribution of inside and outside space on each is different. The inside spaces on the ground floor, arranged around the small courtyards, are mainly lobbies and the occasional bathroom. Living rooms and bedrooms are distributed amongst the upper two floors. Each of the top floor apartments has a roof terrace. On the middle floor there are two small covered terraces, for air.

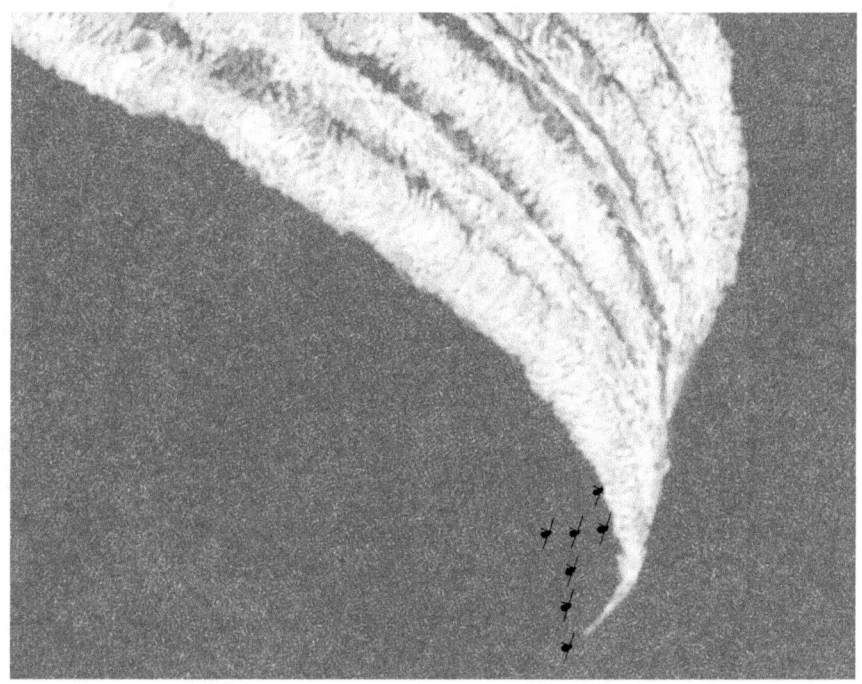

MOVEMENT CURVES

We might try to move in straight lines but when we throw a ball or swing a bat, when we dance or run or do gymnastics… we find that our movements are more about curves than straight lines. Architecture accommodates curves of movement. In the preceding examples, architecture sets an orthogonal frame as an arena for movement. Its rectangularity acts as a foil and counterpoint to the fluid melody of movement. But architecture can respond in different ways to movement. It can relate to movement more directly. Its form can channel and express movement. And in taking us on journeys, architecture tends to become more curved.

CURVES AND MOVEMENT
the route is not always straight

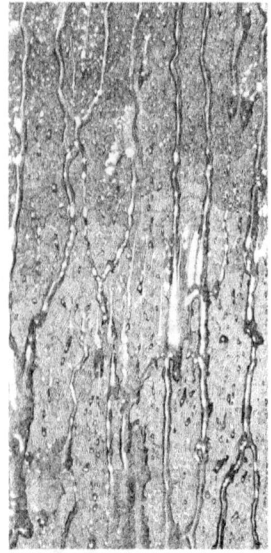 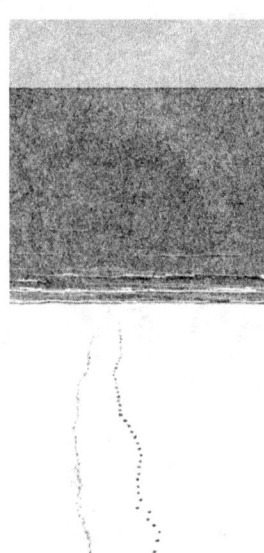 

Pulled by gravity, blown by wind, encountering grease and debris, the shortest route for a raindrop on glass is not a straight line.

On the beach, you and your dog might think you are walking in a straight line but your paths are curved.

Deviated by topography, avoiding obstacles, a pathway through the woods is never a straight line.

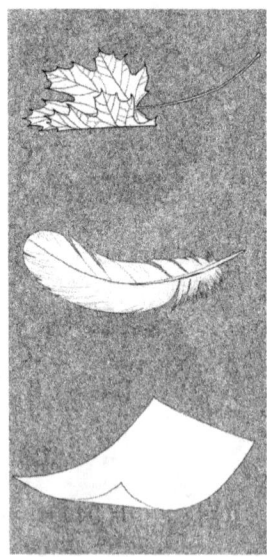  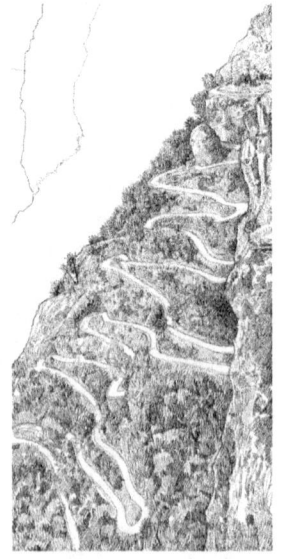

In air, leaves, feathers, paper... do not fall in straight lines.

Snakes wind their way across the ground in wavy curves.

To climb steep slopes roads switch back on themselves in curves.

THE ORTHOGONAL CONSTRAINT
box-like rooms constrict freedom to move in curves

The compartmentalisation of my house into box-like rooms has the effect of constraining my movement to straight lines and right angles (right). The exception that proves the rule is the one instance where a curve is established by the architecture: the pathway from the gateway in the hedge to the front door.

On open land, by contrast, we are free to move and settle where we might. On the beach (below) we can run in circles and loops, unconstrained by walls and doorways. In the sea we can swim or sail where we wish. If we were birds we could fly freely in three dimensions. (Even so, on the beach we sometimes set down for ourselves architectural rules: the rectangular towel or tent; the rectangular games court.)

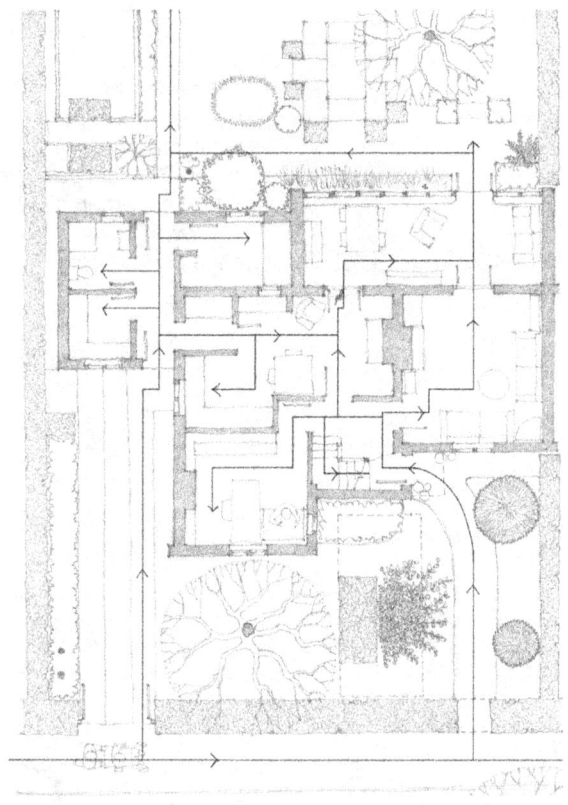

GESTURAL CURVES
graphic curves

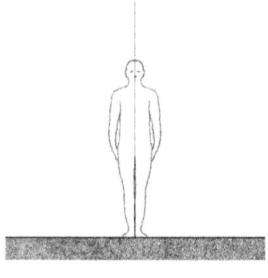

The curves of our movements contrast with the straight line stillness of standing to attention. Our gestures can be recorded graphically, freezing movement in an image. As demonstrated in the work of abstract expressionist painters (such as Jackson Pollock) such frozen records of seemingly random movements can be aesthetically powerful. Their power derives from their combination of chance and control.

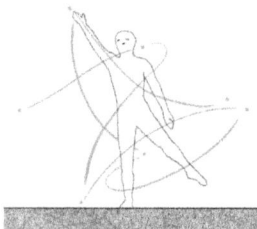  

For example, Rudolf von Laban, in his Gammes Dynamosphériques (Dynamospheric Ranges; some of which are redrawn here), illustrated the reach and arcs of movement of our various limbs with sweeping curves. The illustrations, as well as having a didactic purpose, show how such drawings can be expressive of dynamism and emotion as well as movement.

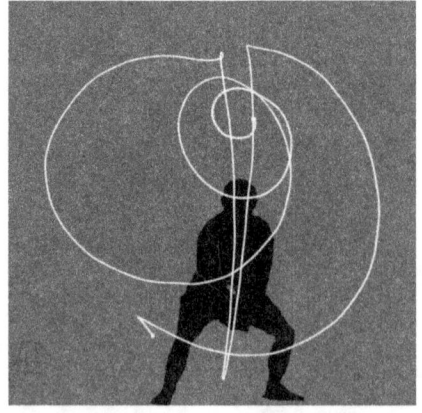

Traditional calligraphy, created with brush movements of varying pace, can be expressive as well as communicative (above). The strokes are produced directly by the muscular gestures of the calligrapher, and produce a quality that blends the calligrapher's nature and mind. They mean more than the words drawn.

In the late 1940s, Pablo Picasso played with light and photography to produce light drawings (above). In a dim room, with the camera shutter on an exposure of a few seconds, he drew in the air with an electric light. Like calligraphy, the images are a record of his spontaneous movements.

EXPRESSIVE (EXPRESSIONIST) CURVES
Einstein Tower, Germany, Erich Mendelsohn, 1921

The Einstein Tower is an expression of Mendelsohn's image of the future as dynamic and curvaceous.

From the images of cities in futurist comics it seems the trajectory of human civilisation is towards the curve.

After the First World War the German architect Erich Mendelsohn designed a tower in Potsdam near Berlin. It was to house equipment, including telescopes, to study the stars, and was appropriately named the Einstein Tower (left). It could be argued its design was informed more by a vision of the curved future than by pragmatic considerations.

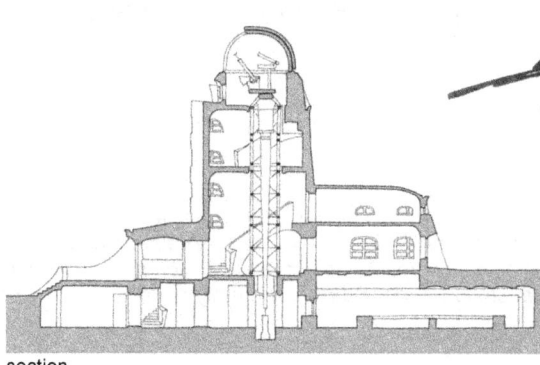

© bpk/Kunstbibliothek, SMB/Dietmar Katz

Mendelsohn's intention is best illustrated by the expressive sketches he made of the tower (above). Almost gestural, these convey the energy he ascribed to the future.

section

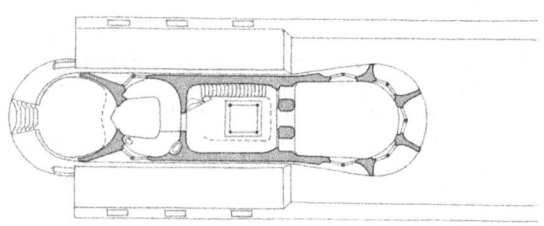

plan of entrance floor

But the curving lines of the rapid brush strokes of Mendelsohn's sketches had to be reconciled with the constraints of actual building. It was not possible to build directly from the sketch. It had first to be converted into conventional architectural drawings (left) which would take into account pragmatics.

TRYING FOR FREEDOM IN ARCHITECTURE
the 'free plan'

In the twentieth century some architects sought to create an architecture that did not constrain movement within rectangular box-like rooms.

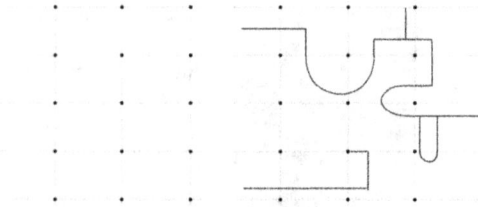

Le Corbusier, for example, promoted the idea of the 'free plan'. It would take advantage of the concrete or steel frame structure to allow walls, free of their need to support floors and roofs, to be placed anywhere.

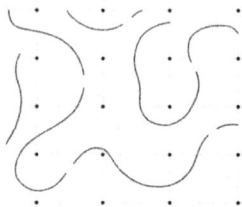

The drawing above (right) is a neatened version of a sketch Le Corbusier did to illustrate the idea of a free plan. Many of the walls do not break free of the orthogonal matrix defined by the rectangular structural grid. But some express their 'freedom' by being curved. Theoretically, a Corbusian free plan could be comprised completely of curving walls (left).

Constrained only by the grid of columns, and therefore seemingly offering open space, frame structures offer spaces in which we can wander freely, in curving pathways too if we wish, as if on an open beach.

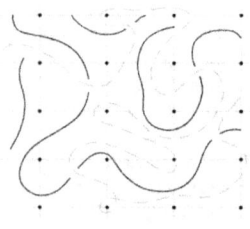

Combining the two aspects of the free plan, operating in the horizontal dimensions, a plan can be produced that combines the freedom of walls to be curved and our own freedom to wander in curved pathways too.

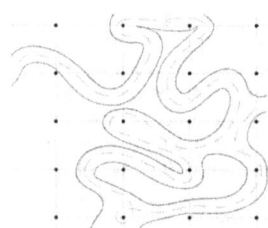

But sometimes the combination of curved walls and curved pathways is far from free. Pathways defined by curving walls can be even more constraining in how they allow us to move than orthogonal arrangements of box-like rooms.

THE MUSIC (AND DANCE) ANALOGY
rectangular rhythm and curving melody

Le Corbusier's mother was a music teacher. It is tempting to interpret his ideas about 'free' plans as an architectural manifestation of music.

Above are the three voice lines of the beginning of a trio in a Mozart opera. All the architecture, the 'spatial' rules, of the music are evident: the clef, indicating the tonal range; the key signature, indicating the key; the time signature indicating the rhythm. The beats represented graphically in the sheet music are numbered above, and regularised in the lower diagram. Through these spatial rules the melodic lines are intertwined. The lower diagram is like the plan of a columned hall through which three individuals follow their own curved pathways whilst remaining in conversation with the others.

An architectural expression of the same idea would be dancers performing in an actual columned hall (below). Their movements, as well as the music, form curved lines within the spatial order of the grid of columns.

'Soave sia il vento,
Tranquilla sia l'onda,
Ed ogni elemento
Benigno risponda
Ai nostri desir.'

'Breeze, be gentle,
Waves, be calm,
And all elements
Respond kindly
To our desire.'

Mozart/da Ponte – *Così Fan Tutte*, 1790.

CURVING PATHWAYS, CURVING WALLS
open world versus curved world

There can be open spaces that allow free curved lines of movement; and there can be spaces defined and occupied by curved forms – walls, ceilings and even floors.

The principle of the bürolandschaft approach to office design (right) was to create large open spaces, perhaps only punctuated by structural columns, in which furniture could be arranged in various permutations, and through which pathways could meander as on a beach.

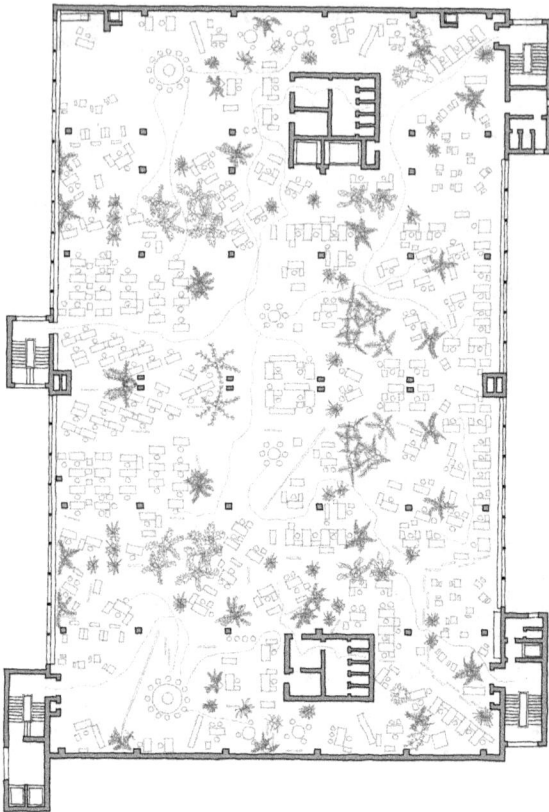

Bertelsmann Verlag, Gütersloh, 1950s
Leonardo Glass Cube, 3Deluxe, 2007

The Leonardo Glass Cube in Bad Driburg, Germany, is for exhibitions, meetings and hospitality. Its glass walls form a square in plan (right) but inside is a composition of elements that curve in all three dimensions incorporating sitting areas, stairways and display spaces. Outside, the curves extend across the surrounding grass as a network of curving 'pathways'.

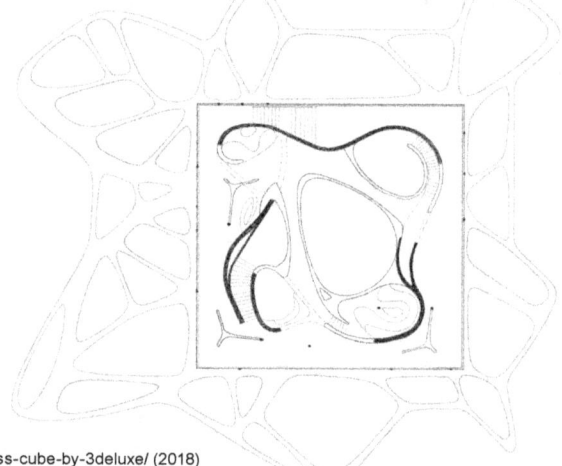

dezeen.com/2008/08/28/leonardo-glass-cube-by-3deluxe/ (2018)

ANALYSING ARCHITECTURE NOTEBOOKS

CURVING FLOORS
Bioscleave House, Madeline Gins and Arakawa, 2008

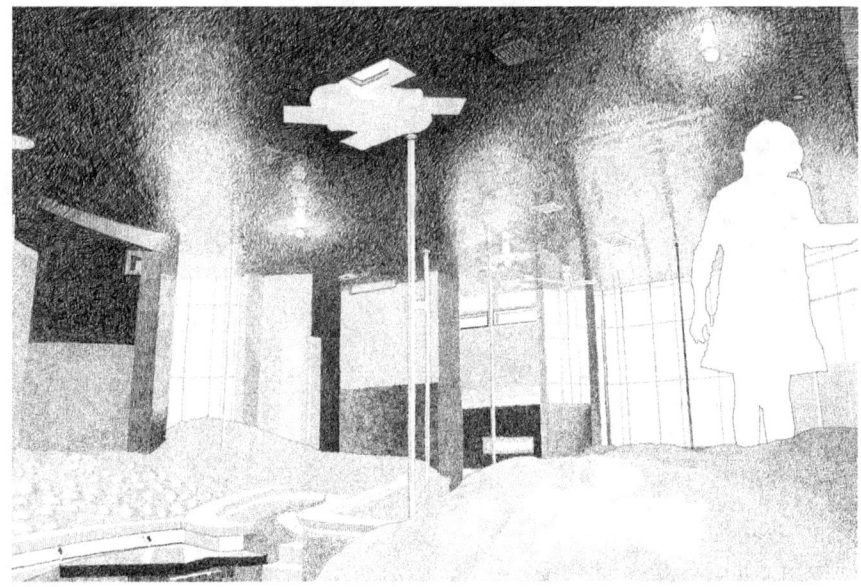

interior

Some architects have even experimented with curvy floors. The Biosclave House* in East Hampton, New York, has curved walls, a sloping ceiling, and a hillocky floor, like uneven natural terrain. The intention was that clambering up and down would keep occupants fit, and help them live forever.

* Simon Unwin – *Twenty-Five Buildings Every Architect Should Understand*, 2015, pp. 255–64.

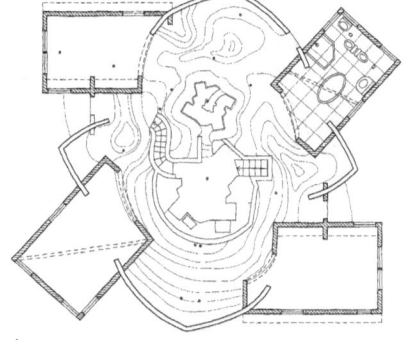

plan

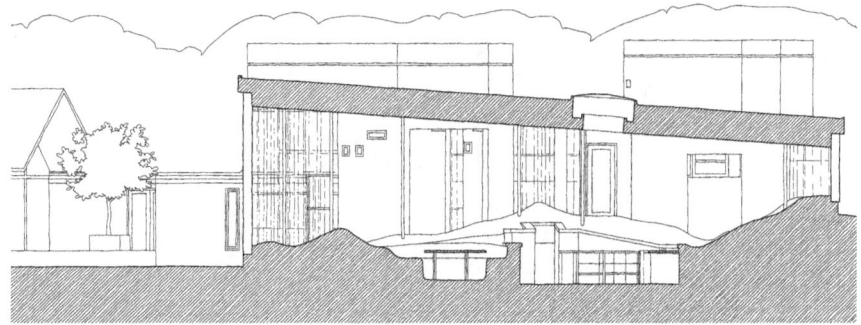

section (not to the same scale as the plan above)

CURVING PATHWAYS
exploring a garden or retail opportunity

The curving routes of a traditional Japanese stroll garden (right) channel our movement along specific pathways. The architect's intention is to lead you through a series of aesthetic interactions with the garden, which change as you move, at different times of day, and with the seasons.

The defined curving pathway of a stroll garden may be laid out with benevolent intent, but that of an airport retail zone might have a different purpose. The one below, which wends between the structural columns of Stansted Airport, and which is lined with retail outlets, lengthens your route from security to gate, not so you can enjoy the extra walking but so that sales opportunities are optimised.

Hakusasonsu Garden, Kyoto

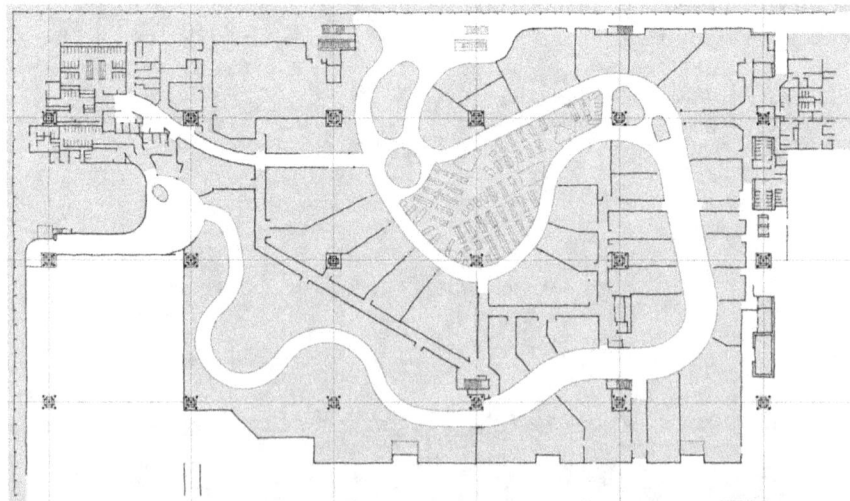

part plan of Stansted Airport, UK (Foster+Partners, 1991; retail layout, World Duty Free Group, 2014)

CURVES FRAMING MOVEMENT
curved routes for aesthetics and economy of space

Pathways – lines of movement – might be curved for a variety of reasons. On page 42 there are examples where the curve of a pathway is caused by obstacles – the trees of a forest – or prompted by the need for an ingenious response to a problem – the desire to construct a road up a steep hillside.

Pathways might also be curved to enhance aesthetic effect and the experience of those who pass along them. The line of stepping stones below is in another traditional Japanese garden. It is gently curved so the subtle composition presented to the eye is not compromised by a straight line, as well as to help make you think carefully about where you are placing your feet as you cross.

The shortest, or the 'best', route between two points is not necessarily a straight line. pathways can be curved for practical reasons too. A spiral stair accomplishes the shortest route between two points vertically and within the smallest possible space (without using a mechanical lift). One of the earliest is that inside Trajan's column (right).

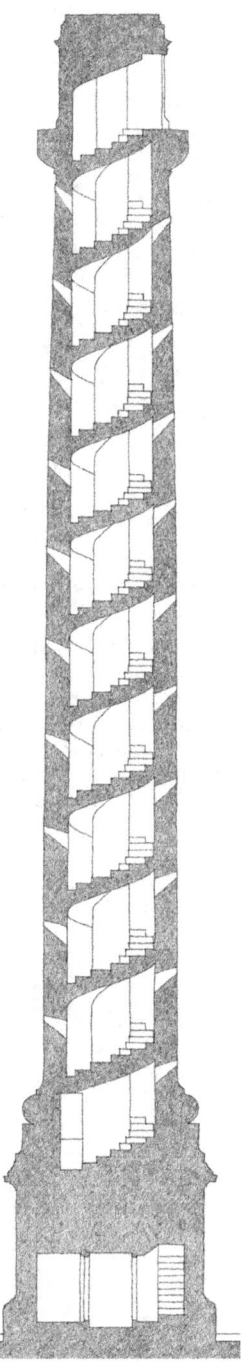

These stepping stones curve gently across the stream.

Trajan's Column, Rome, 113 CE

CURVING THROUGH THREE DIMENSIONS
swooping through space

Architecture accommodates movement in space and time. Curved routes can enhance our experience of moving through space. They offer changing views. They offer drama and disorientation.

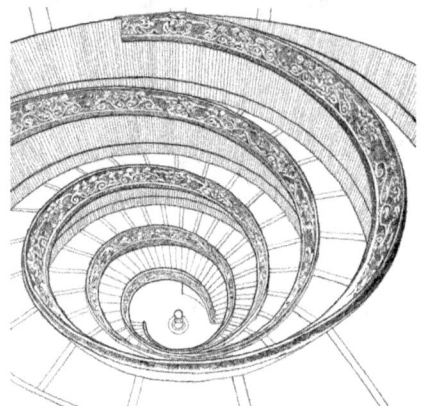

Spiral ramp, Vatican, Giuseppe Momo, 1932

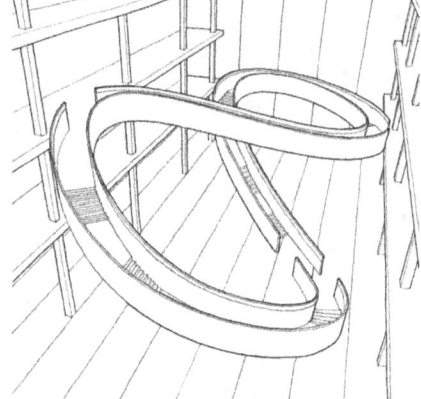

'infinity' ramp, 4A Four Architects, 2014

Curving stairs and ramps can look sculptural – they are three-dimensional forms that appeal to the eye. They also engage our sensibilities and emotions by providing us with a special kind of experience; they provide a pathway for movement through all three spatial dimensions and the fourth dimension of time. Such movement might be sedate and reflective but it can also be dramatic and scary, as on a roller coaster.

INFINITE CURVES
Endless House, Friedrich Kiesler, 1947–61

Friedrich Kiesler's design for an Endless House (above and right) was based on a scribble – a record of movement rather than the depiction of an object. Scribbles tend to be tangles of curving lines, looping and spiralling in the illusory space of a two-dimensional sheet of paper.

Kiesler wanted his house to be a container for endless looping and spiralling movement too. And the endlessness of the pathways around, into and within the house are curves weaving through the four dimensions of space-time.

* Simon Unwin – *Twenty-Five Buildings Every Architect Should Understand*, 2015, pp. 51–62.

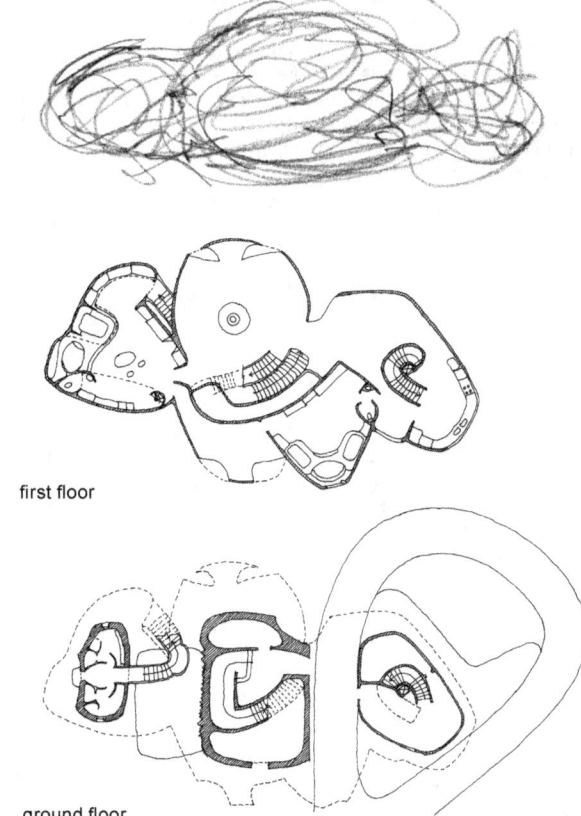

first floor

ground floor

CURVE

ARCHITECTURAL PROMENADE
Le Corbusier and Sverre Fehn

Making pathways for movement has always been part of architecture. Pathways wandering people and animals make in the landscape, merely by the wear or even in the memory of repeated passage, are part of the architecture we introduce into the world. Many great buildings, whether temples or churches, establish pathways along doorway axes – pathways of emergence of or aspiration.

In the twentieth century, Le Corbusier promulgated the idea that the pathways defined by architecture could be for wandering. He called this idea the 'architectural promenade'. Such exploratory routes could (should) be curved rather than axial. The idea is apparent in his Maison La Roche (1923; below) with its ramp curving up from the picture gallery to the library. But the most celebrated example is the Villa Savoye (1929; right), where the route begins with the driveway curving under the overhanging first floor and continues up through the building to the roof.

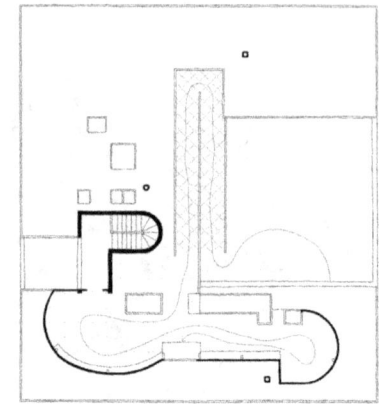
Villa Savoye, roof plan

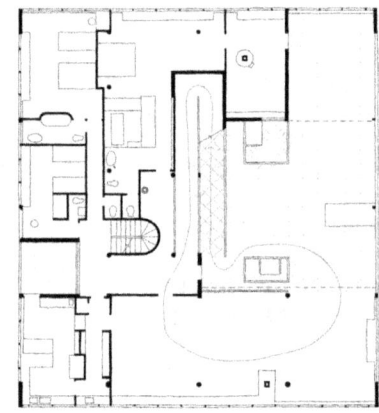
Villa Savoye, first floor plan

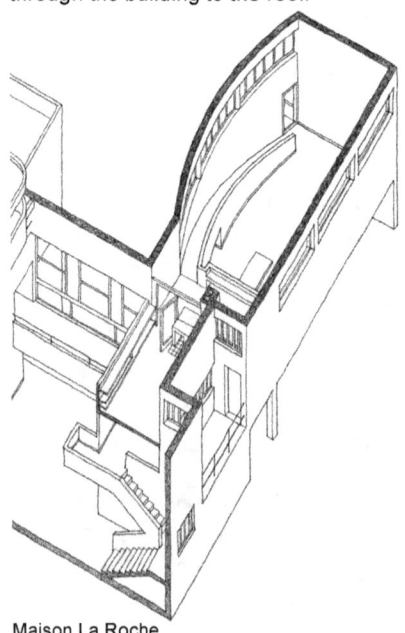
Maison La Roche

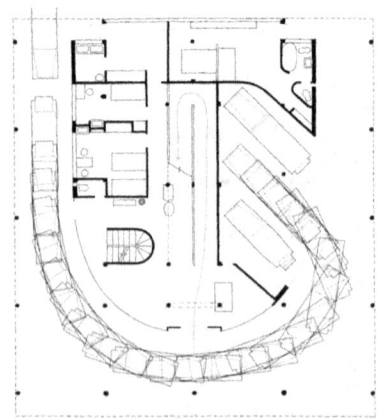
Villa Savoye, ground floor plan

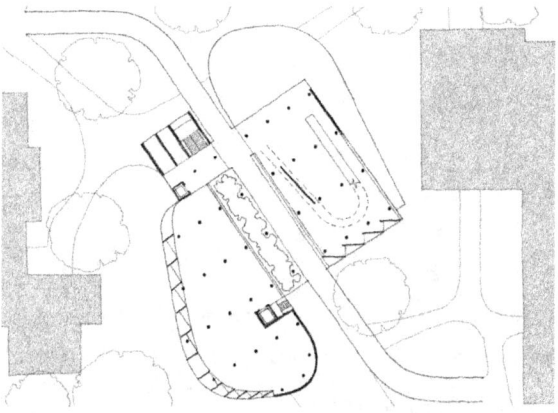

The ramps that curve up towards, into and through Le Corbusier's Carpenter Center for the Visual Arts (left; 1964) extend visitors' experience of entering and give them a changing view of the building as they approach.

Le Corbusier used the idea of the architectural promenade in many of his works and throughout his career. The Carpenter Center for the Visual Arts in Cambridge MA (above), built some forty years after the Maison La Roche, has a ramp that curves up and through the building.

Other architects have used similar devices. At the Hedmark Museum in Norway, the architect Sverre Fehn converted an old barn built over the remains of a medieval bishop's palace into a museum (below). A ramp curves up over the layers of history exposed below, giving the visitor varying views and extending the process of entrance.

According to Feng Shui, the ancient Chinese philosophy for harmonising people with place, it is more auspicious that the approach to a building is curved rather than straight.

The entrance ramp at the Hedmark Museum in Hamar (below) curves up over the layers of archaeology before entering the building at the top of the auditorium. As in the case of the Carpenter Center above, the curve in the ramp enhances the experience of approaching the building.

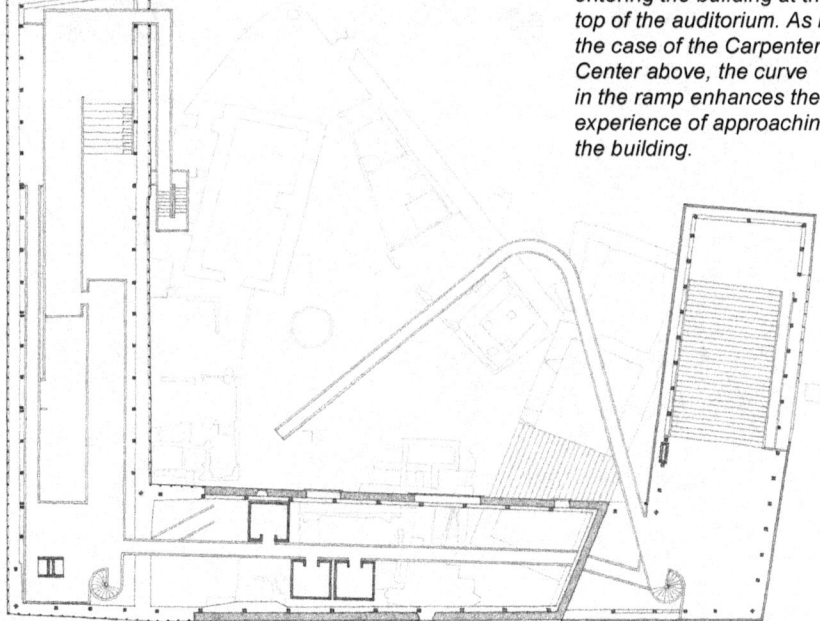

Hedmark Museum, Hamar, Norway, 1967-79, upper-level plan

CREATIVE ASPIRATION
Guggenheim Museum, Frank Lloyd Wright, 1943–59

Through his long career, the vast majority of Frank Lloyd Wight's buildings, and arguably all his best work, had been orthogonal or based on geometrical compositions of straight lines. But in the 1940s he produced a design for a building thoroughly informed by curves. It was the first Guggenheim Museum, completed in New York a few months after Wright's death in 1959.

By comparison with some of the architectural curves of the twenty-first century, the curves of the Guggenheim are restrained. It seems Wright could not completely free himself from his belief in the authority of Platonic geometry – squares, circles, pyramids, cones…

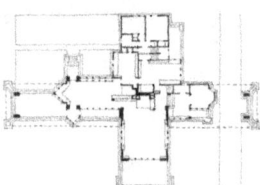

Ward Willits House, 1901

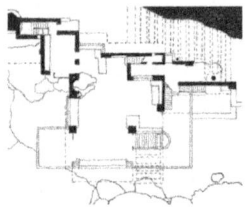

Fallingwater, 1935

With a few exceptions, through his long career, Wright's designs had been governed by the rectangle, straight lines and a square grid.

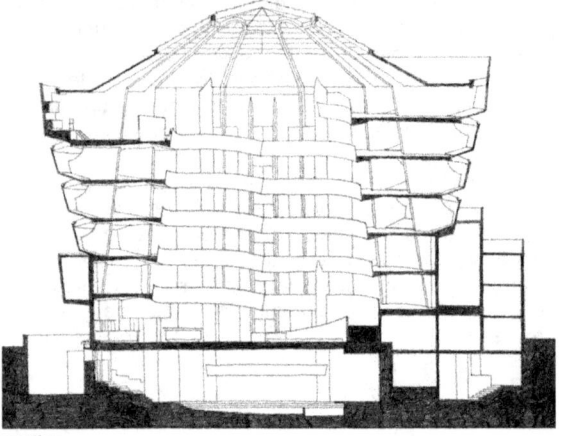

section

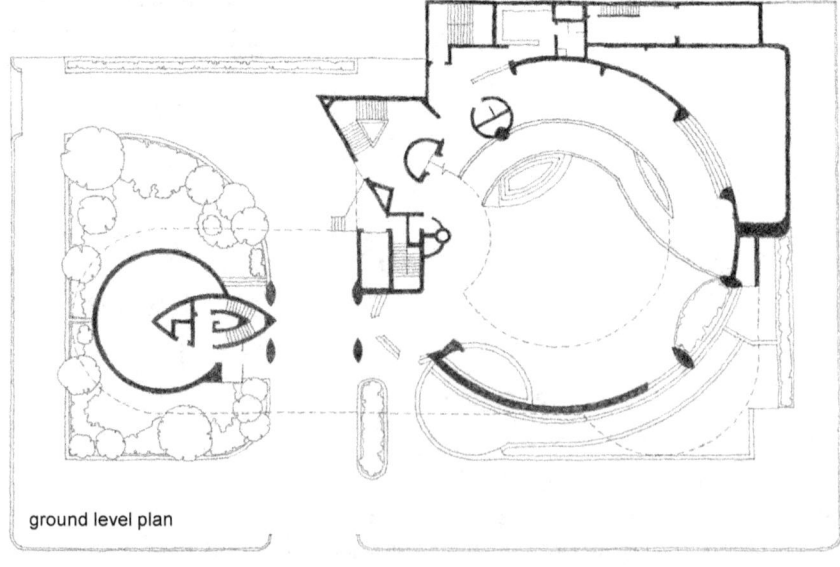

ground level plan

The curved geometry of the Guggenheim contrasts with the regular grid of Manhattan and the rectangular buildings that cram into its blocks (right). The circular forms of Wright's design, like sculpture, demand space around them. A round peg in a square hole leaves space in the corners of the hole; the convexity of circles will not allow them to jigsaw neatly together like orthogonal blocks. Some might say Wright's curved geometry is profligate in a city where space is at a premium and expensive, and inappropriate in a city of skyscrapers. But, adjacent to Central Park, Wright's sculptural building stands out as an ornament in a way that a rectangular block building could not. As a showman, and like some other architects, Wright would have enjoyed the attention his building demanded and which it attracted mainly because of its idiosyncratic curved geometry and off-white windowless surfaces. (Originally Wright wanted the Guggenheim clad in red marble.)

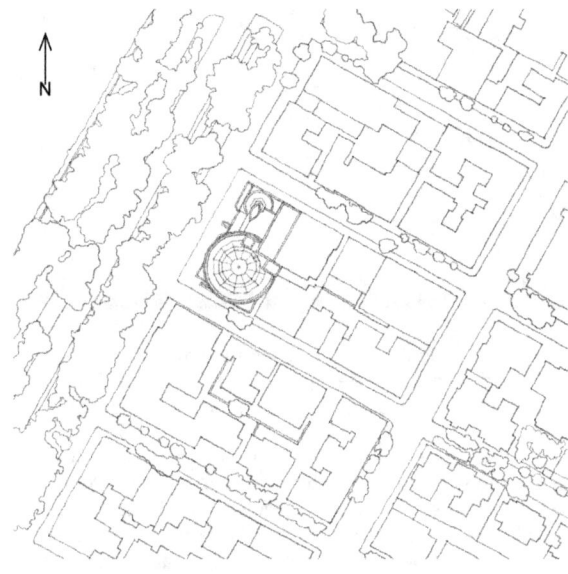

Wright tried spiral ramps in previous projects. His 1924 designs for an Automobile Objective on Sugarloaf Mountain, Maryland, had a double ramp. The 'objective' was that people could drive to the top to enjoy the view. One of the ramps (the upper) was for them to drive up, and the other (below it) to drive back down. Inside was going to be a large domed planetarium. The project was not realised, but obviously the idea of a ramp stuck in Wright's mind, re-emerging twenty years later.

Wright may have had historical precedents in mind, such as the minaret of the Great Mosque of Samara, Iraq (ninth century CE).

Wright's first design for the Guggenheim (above) seems to have drawn on his design for the Automobile Objective from twenty years earlier. It has the appearance of a ramp on the exterior, but here the external form expresses the presence of an internal spiral ramp for visitors to walk up admiring the art works hung on the inner surfaces of the outer walls, washed by light from the sky. This design was intended for an unspecified site in downtown New York, in a different situation from the site eventually used, which was not acquired until 1949. But even in this early design it is clear that Wright wanted to counter the unremitting rectilinearity of Manhattan with a curved sculptural form.

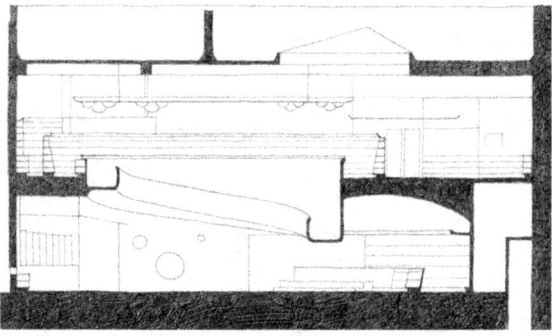

section

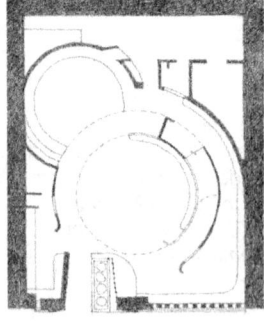

ground floor plan

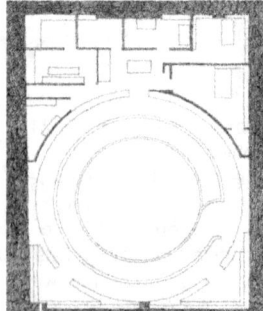

upper floor plan

In 1948, after the very earliest designs for the Guggenheim, Wright was engaged to design the façade and interior of the V.C. Morris store in San Francisco (plans and section, below left). Here he counterpointed the internal rectangular box-like space available with a circular ramp rising one floor. The central portion of the shop, encircled by the ramp, was lit from the sky.

In the first designs for the Guggenheim and in the design for the V.C. Morris store we can see the principal elements of the eventual Guggenheim design: an internal spiral ramp encircling a central space lit from the sky; the expression of that ramp externally; and light from the sky washing the walls on which art work was to be hung.

The V.C. Morris Store, now the Xanadu Gallery, at 140 Maiden Lane, San Francisco, is one of the buildings you can 'walk around' on Google Earth.

These drawings are based on published drawings and are not an exact depiction of how the store is today.

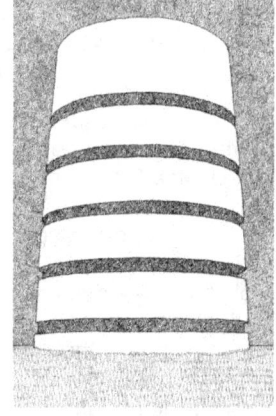

The primary curve of Wright's design for the Guggenheim is an upward spiral of movement. This may be read as a metaphor for aspiration. Like the earlier Automobile Objective project the museum has an outwardly visible expanding spiral that gives the building its sculptural and expressive appearance (above). The inner spiral decreases in diameter as it rises, making the central space appear even more lofty than it is. But the main achievement of the building is that it takes the visitor on an ascending journey, along the pathway between the outer and inner spirals.

The Guggenheim takes advantage of the smooth curved monolithic forms made possible by reinforced concrete technology, as also used in defensive structures in Europe during the Second World War. (Above is one of the German defensive structures on the Channel Island of Guernsey). The museum's purpose was more benign, but it benefited from the futuristic imagery produced by using this technology, which allowed cantilevers and sloping walls as well as smooth surfaced curves. (Though, like Mendelsohn's Einstein Tower on page 45, in the twenty-first century it looks more like 'retro-futurism'.)

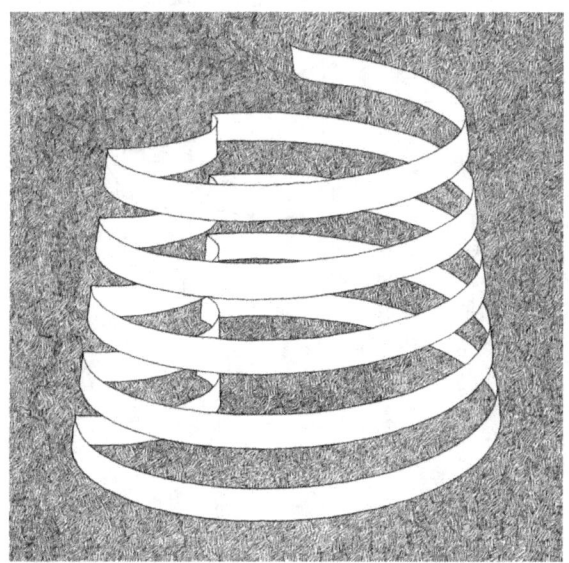

It is also possible to take the lift (elevator) to the top of the spiral and make the journey, viewing the art, downwards instead of upwards.

CURVE

SPATIAL PIROUETTE
Truss Wall House, Ushida Findlay, 1993

You might remember (from a previous book*) that Ushida Findlay's Truss Wall House in Japan (1993) can also be interpreted as the realisation of a dynamic squiggle in built form. It too is a house that frames a pathway – from street, through living spaces and a small courtyard, up a stair to a roof terrace and seat – which curves in an irregular S-shaped spiral like a gestural flourish or a dancer's pirouette.

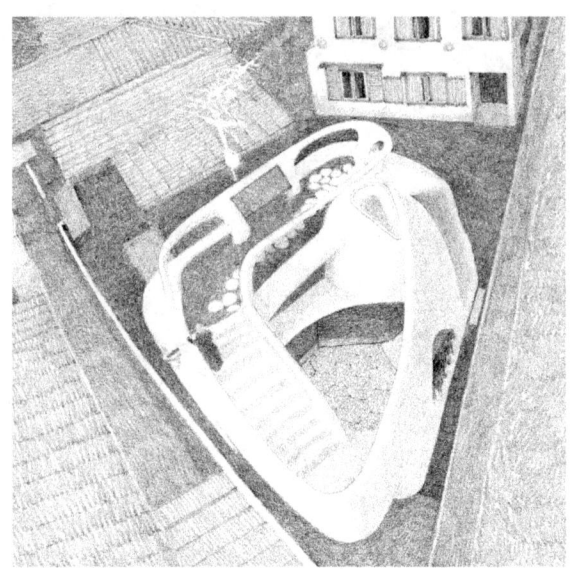

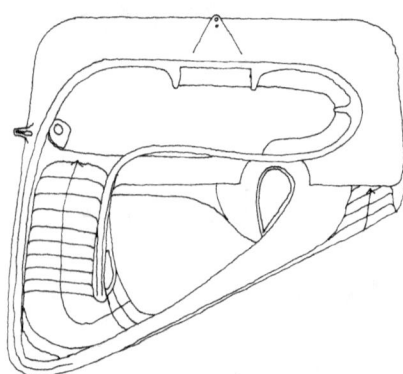

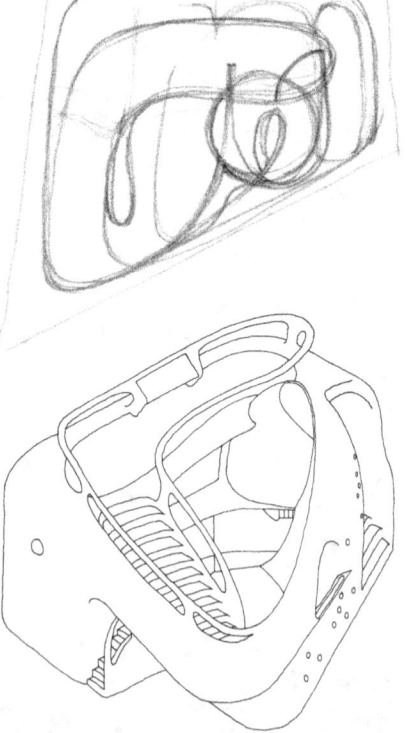

In both Kiesler's Endless House (page 53) and the Truss Wall House, curves are not only an expression of movement in static sculptural form; they also define pathways – lines of actual movement to be followed by their inhabitants. These are buildings that make people perform.

* Simon Unwin – *Twenty-Five Buildings Every Architect Should Understand*, 2015, pp. 43–50.

STRUCTURAL CURVES

Considering their association with deceit, perhaps it is understandable that curves – in their contributions to architectural structures – might be considered the equivalent of magic spells. Curves can make possible constructions that otherwise would be unstable. Arches and vaults may span spaces much larger than can be roofed with timber beams or stone lintels. There are many types of structure that depend on curves but the factors underlying the generation of those curves can be very different. Traditionally we have had a preference for generating structural curves using mechanical means rather than mathematical calculation. Practical builders would rather work with pegs and string than with slide rules or calculators.

STRUCTURAL LIMITATIONS OF STRAIGHT LINES
straight stones or timbers will only span so far

In accordance with the principles of orthogonality outlined in the first chapter of this Notebook, buildings are often composed of straight lines and vertical elements. But the spans that can be achieved by horizontal elements are limited. Because of its relative weakness in tension, stone will crack if asked to span more than a short distance, as evident in Greek temples (left).

A classical Greek temple is composed of vertical walls and columns, with horizontal beams spanning between them. You can see the entablature spanning across the columns in the above drawing, as well as the shadows of the stone beams spanning from the entablature to the cella wall. Both spans are limited by the intrinsic nature of stone.

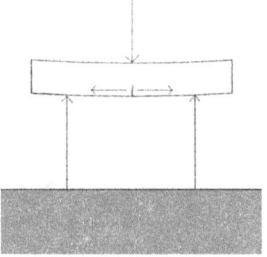

The tension at the bottom of a stone beam or lintel under load will cause it to crack and fail.

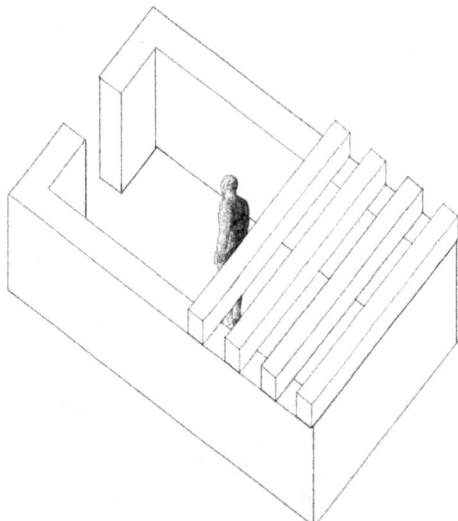

This is the simplest version of a structural strategy used in the vast majority of buildings around the world. The straight lines and right angles of a rectangular room (above right) mean that structural members can be cut to the same length and section to span between parallel walls, making construction easier and quicker. It depends on the sensible geometry of a rectangular room, and spans are limited by the lengths of timber that can be cut from trees.

Because of their strength in tension, timbers will span slightly further than stone. But even so, the span that can be achieved with timber beams is around 6 or 7 metres. If you want to span a larger space, ingenuity is required.

ROUNDHOUSE
an early curved structure

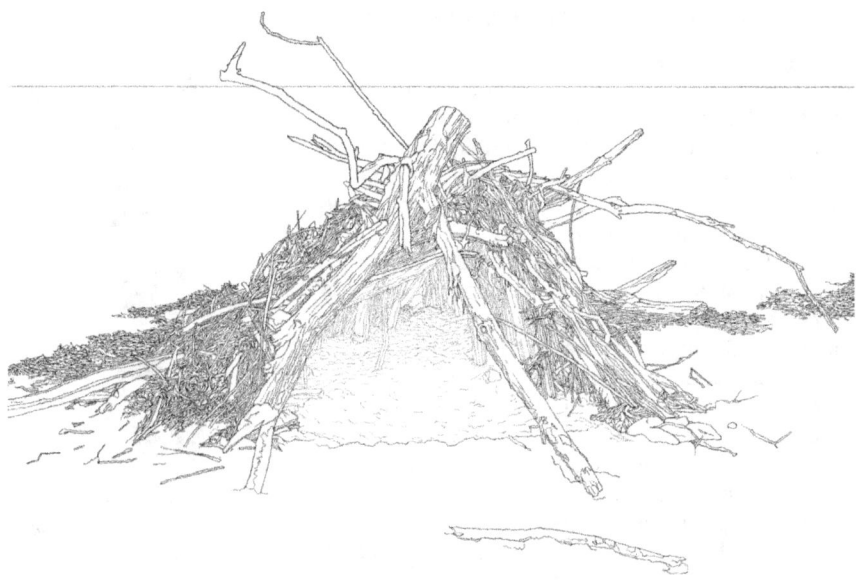

The concept of the round building is probably older than that of the orthogonal. One of the most primitive types of shelter, which we still build today, is a structure built from branches leaning against each other. We do this with driftwood on the beach (above) and it is the technique used by scouts when making a bivouac. The structure naturally creates a conical form, the circular plan of which is in accord with our sense that our own place and that of our family is a circle.

This technique for forming a conical structure was, in history, probably refined before the greater flexibility of orthogonal structures were explored. Archaeological remains of Iron Age houses suggest they were round with conical roofs. The one illustrated below is a twentieth-century interpretation of how such roofs might have been built. It was built at the St Fagans National Museum of History near Cardiff in south Wales. The curved roof was covered with thatch.

CRUCKS
firmness, commodity, delight

'Well building hath three Conditions. Commoditie, Firmenes, and Delight.'

Henry Wotton – *The Elements of Architecture*, 1624.

The cruck frame is an example where a curve is beneficial. It is a structural strategy found in traditional timber-framed buildings, especially houses and barns. It allows a combination of both firmness and commodity (of the three Conditions listed by Henry Wotton above) and can be refined to include the third – delight – too.

    A cross-section through a cruck-framed building is shown in the drawings below. The crucks are the two diagonal timbers (a) resting on the sills and supporting the ridge (b). The crucks are the principal structural timbers. The other timbers are supplementary: the ties (c) hold the crucks together and stop them spreading apart under the load of the roof; additional timbers (d) may be necessary to even out the slope of the roof and help support the purlins (e); the purlins and wall plates (f) support the rafters (g) which in turn support battens (h) to which the roof tiles, slates or slabs (not shown) are fixed; vertical wall frames (i) are fixed between the sill and the wall plate.

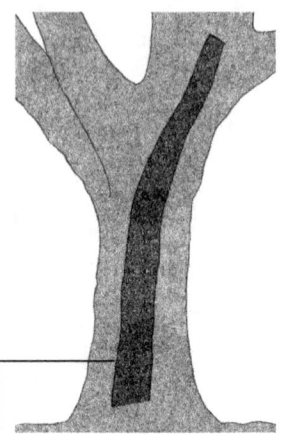

A pair of crucks could be cut from a bent oak, split so that the crucks would be roughly mirrored in the final structural frame.

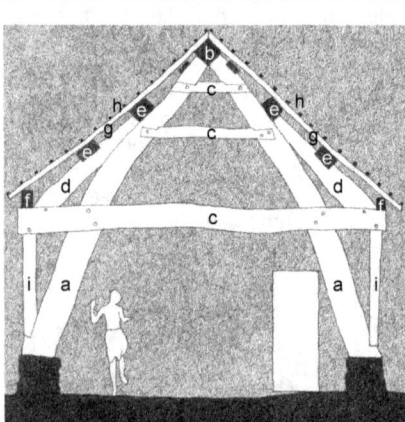

The cruck frame is a strategy in which natural curves suit an architectural aim. This example shows crucks with a moderate curve. A tree with a curve is selected, and the pair of matched crucks made by splitting it in two. Of course different trees will have different amounts of curve and so the overall system has to be able to accommodate variations.

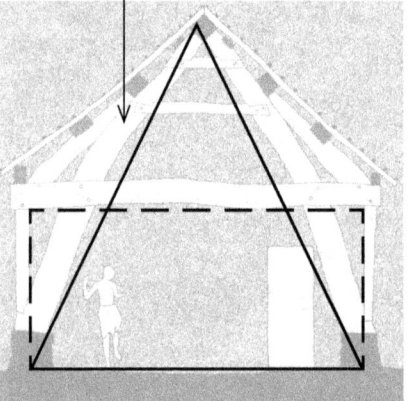

The cruck frame idea is a compromise between two requirements: a rectangular box of space, within which to move around and keep possessions, and structural stability. The latter is provided by the triangular form of the crucks. The rectangular box would not be strong in itself, having a tendency to collapse sideways.

The cruck frame strategy offers a compromise between the spatial desire for commodity and the need for structural firmness. This compromise is made possible by the curve in the timber crucks. Builders found ways of introducing the third of Wotton's 'Conditions' too – delight.

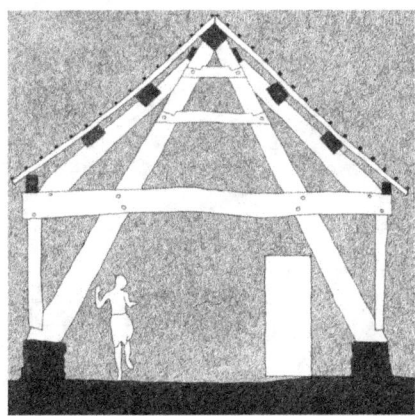

If the tree from which the crucks were cut was straight then the frame would be diagonal and strong. But the crucks would intrude into the rectangular box of living space. You might hit your head on them, and furniture could not be placed out of the way up against the walls. The compromise between space and structure is not perfect.

It is thought that cruck-framed buildings are descended from buildings with triangular structural frames (above); no walls, all roof. But builders found they could increase usable space by differentiating roof and wall, and constructing vertical walls separate from the slope of the structural frame (as shown in the drawings opposite).

Builders could get even better results if they were able to find (or grow specially) trees with sufficient bend to form vertical walls and pitched roofs. This meant many of the supplementary timbers could be dispensed with, allowing more lofty spaces (with no danger of head banging) as well as the placing of furniture out of the way against the walls.

The cruck frame strategy is a compromise between the desire for the commodity of a free rectangular box of space in which to live, and the firmness of a triangular frame. But as builders grew in confidence and aspiration they also introduced the element of delight, creating elegant (as well as practical) curves and decorated bracing.

HORIZONTAL ARCH (OR CORBEL DOME)
Treasury of Atreus, Mycenae, c.1250 BCE

section (Treasury of Atreus)

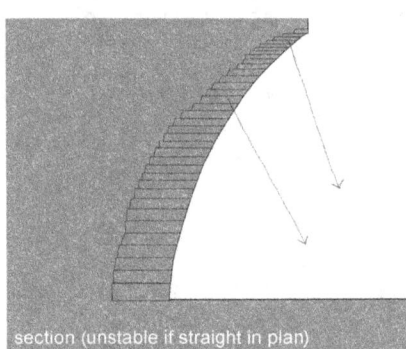
section (unstable if straight in plan)

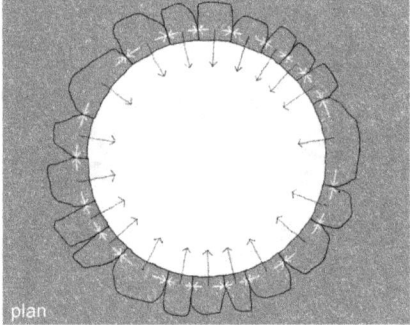
plan

A key to spanning larger spaces with stone was the discovery of the power of the curve, strongest in the form of a circle. The Treasury of Atreus is an example of the corbel dome structure which relies on the stability of the horizontal arch. If a single straight wall curved over, as in the drawing directly above, it would collapse. But when turned into a circle, as in the Treasury of Atreus, it becomes stable. This corbel dome spans a space approaching 15 metres (48' 6"). As shown in the plan, although the stones want to fall inwards, they are prevented from doing so by their neighbours. This is the principle of the horizontal arch.

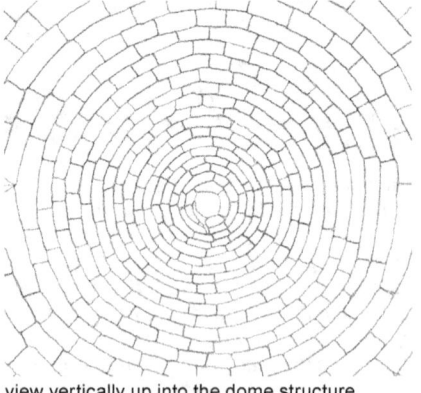
view vertically up into the dome structure

VERTICAL ARCH
a flash of inspiration

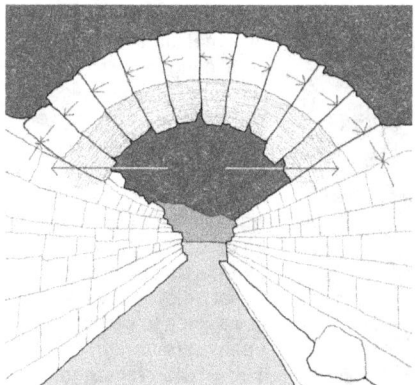

arched vault in ancient Olympia, c.200 BCE

Someone somewhere realised the forces pent up in the horizontal arch could be turned through 90° to create the arch.

Y Garreg Fawr, St Fagans

The vertical curve of the arch transfers load through each voussoir to the walls. Horizontal thrust also has to be resisted.

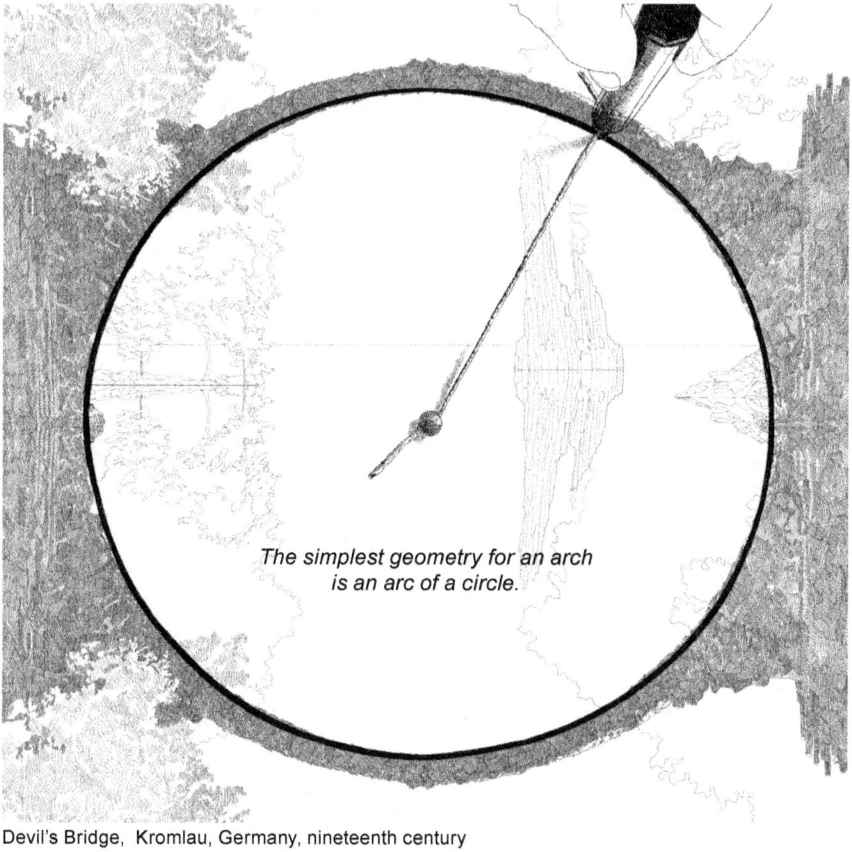

The simplest geometry for an arch is an arc of a circle.

Devil's Bridge, Kromlau, Germany, nineteenth century

SHAPES OF ARCHES
a matter of pragmatics, style and identity

open arches at Pergamon, now in Turkey

Curves always present pragmatists with conflicting issues. The Romans, for example, were aware of the power of the arch and used it frequently to span openings. But when they wanted to be able to close up an opening from time to time, as in the food outlet below, they used flat arches to make the opening rectangular. This was so rectangular shutters could be easily fitted when the shop closed. Curved arches were used to create the overhang sheltering the outside space where customers could sit.

more complex arches at Ostia near Rome

VAULTS
arches in three dimensions

Arches span openings in walls; i.e. conceptually in two dimensions. But arches can be developed into vaults (below) to span over spaces.

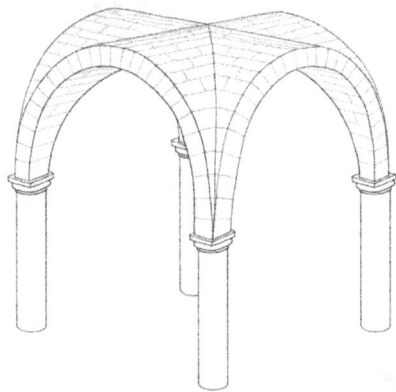

Fergusson – *Handbook of Architecture*, 1859.

Stone vaults were developed in the ancient architecture of Greece and Rome. But their use became more sophisticated in Moorish and Gothic architecture in the Middle Ages. The space in front of the mihrab in the Mezquita of Córdoba (961 CE; top right; see also pages 150–51) has a vault composed of intersecting arches.

Gothic architects made vaults even more complicated (bottom left). But perhaps the most sophisticated vaulted ceiling is that of King's College Chapel in Cambridge (1512–15 CE; below; see also page 161). It is composed of an intricate lattice of many curves, in all three dimensions.

Guadet – *Éléments et théorie de l'architecture*, Volume 3, 1894.

Fergusson – *Modern Architecture*, 1862.

DOME
revolving the arch through 360°

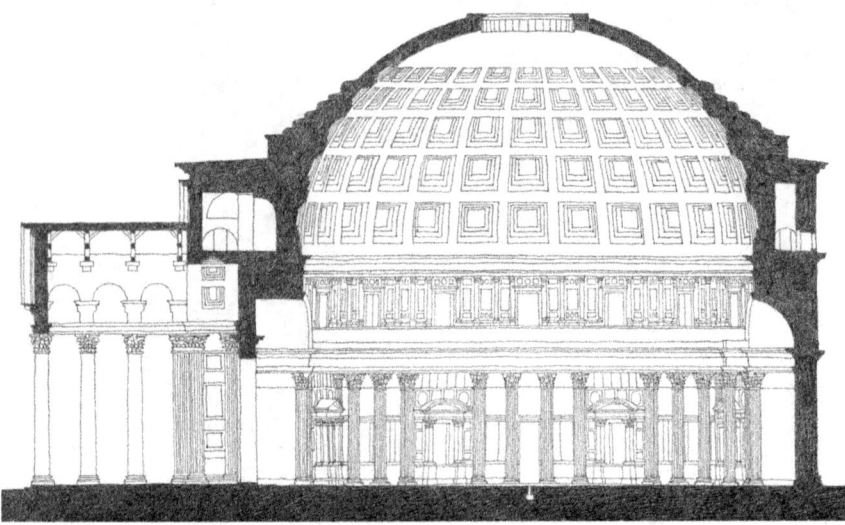

The principle of the arch — by which weight is transferred along curved paths to the ground — also allows the construction of domes, in which the arch is revolved horizontally through 360°. The dome of the Pantheon (above) is monolithic, constructed in mass concrete. Structurally it operates like the shell of an egg or the skull's cranium.

The dome of Hagia Sophia (below) was built of bricks and mortar, with ribs (visible below) to strengthen it (as in some shells, opposite). The outward thrust of the dome is resisted in all directions by half domes built against its sides.

The diameter of the dome of the Pantheon in Rome, built in the 120s CE, is about 43 metres (142 feet).

The diameter of the dome of Hagia Sophia in Istanbul is 31 metres (102 feet). It was built in the sixth century CE.

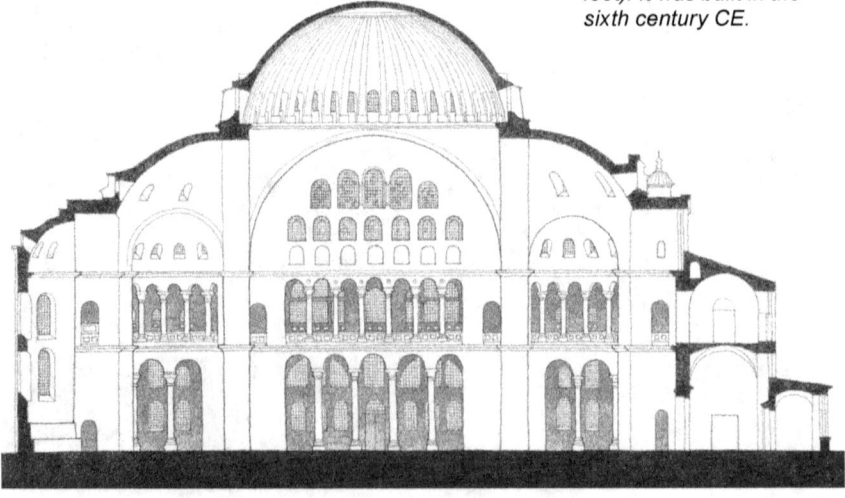

SKULLS AND SHELLS
models for architectural structure

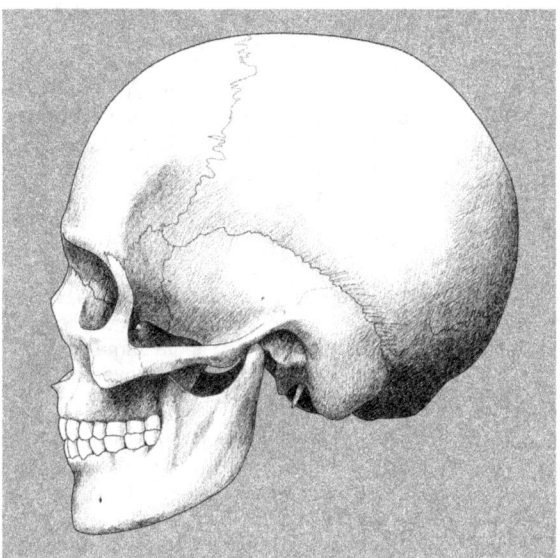

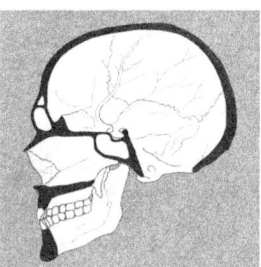

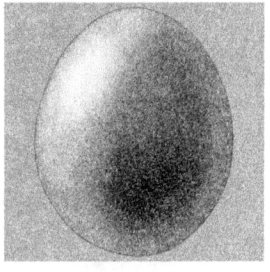

The dome of the Pantheon was cast in concrete on elaborate formwork, subsequently removed. Once set, the concrete formed a single domical entity, like the bone dome of the skull or the shell of an egg, both of which, despite being light, are strong. Compressed along its long axis an egg is difficult to break because the forces are distributed longitudinally through its circular form. The same happens in the dome of the Pantheon.

The thin shells of skulls and eggs are strong because of their curves. (See also the concrete roofs of the Brynmawr Rubber Factory on page 84.)

Sea shells are strong too. This example is strengthened by its ribs, a feature shared by the dome of Hagia Sophia, which was constructed of brick and mortar rather than concrete, but also constructed on elaborate formwork.

LEANING (THREE-CENTRED) ARCH
New Gournia, Egypt, Hassan Fathy, 1940s

To build a vertical arch you first need to build temporary centring to support the stones until the arch is complete.

By building upwards in stone rings of gradually decreasing diameter, it is possible to build a corbel dome (page 66) without temporary support. This is because of the strength of the horizontal arch principle, which does not depend for stability on weight pressing down on it from above.

Building vertical arches is more tricky because they are not strong until complete. All the voussoirs have to work together. Because of this it is necessary to build a temporary framework called centring (left), which is the same shape as the final arch. This is removed once the arch is finished.

The need for centring makes building arches and arched vaults more laborious. In the mid-twentieth century the Egyptian architect, Hassan Fathy, who was interested in producing lost-cost housing according to traditional local principles, turned to ancient precedent to find a method of building vaulted roofs without the need for centring.

The process began by describing an approximately parabolic or catenary outline on an end wall, using the method shown in the drawings on the left. Then light mud brickwork was built up in sloping courses leaning against the end wall (below). Gradually, the vaulting progressed over the space. Examples of this technique have lasted centuries.

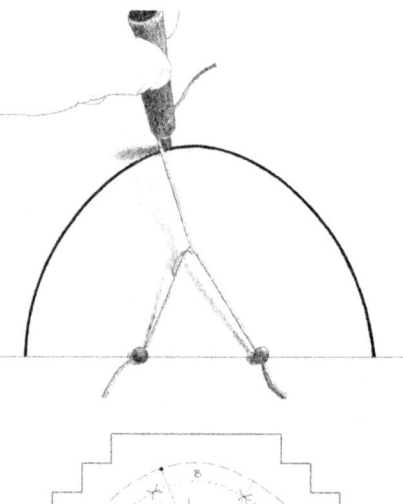

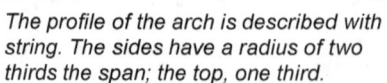

The profile of the arch is described with string. The sides have a radius of two thirds the span; the top, one third.

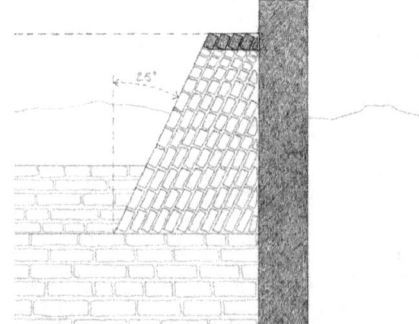

Thin courses of mud brick are laid at an approximate angle of 25°, leaning against the end wall.

THREE-CENTRED ARCH
Tāq Kasrā (Archway of Ctesiphon), Iraq, 6thC CE

Some grand arches and vaults have been constructed using the method (described opposite) adopted by the twentieth-century architect Hassan Fathy. The grandest is some 1500 years old. The Tāq Kasrā, or Archway of Ctesiphon, has crumbled at the edges but the main part remains suspended up to 37 metres above the ground. It was constructed without the need for centring by leaning successive courses against an end wall (see opposite). But scaffolding would have been needed for the builders to reach their work.

The shape of the Ctesiphon Arch seems to match the method of geometric construction also used by Hassan Fathy's workmen some 1500 years later. Though it might look like a parabola or catenary curve (see the following pages), it is constructed of three arcs of circles (see opposite).

CURVE

POINTED ARCH
Al-Aqsa Mosque, Jerusalem, 8ᵗʰC CE

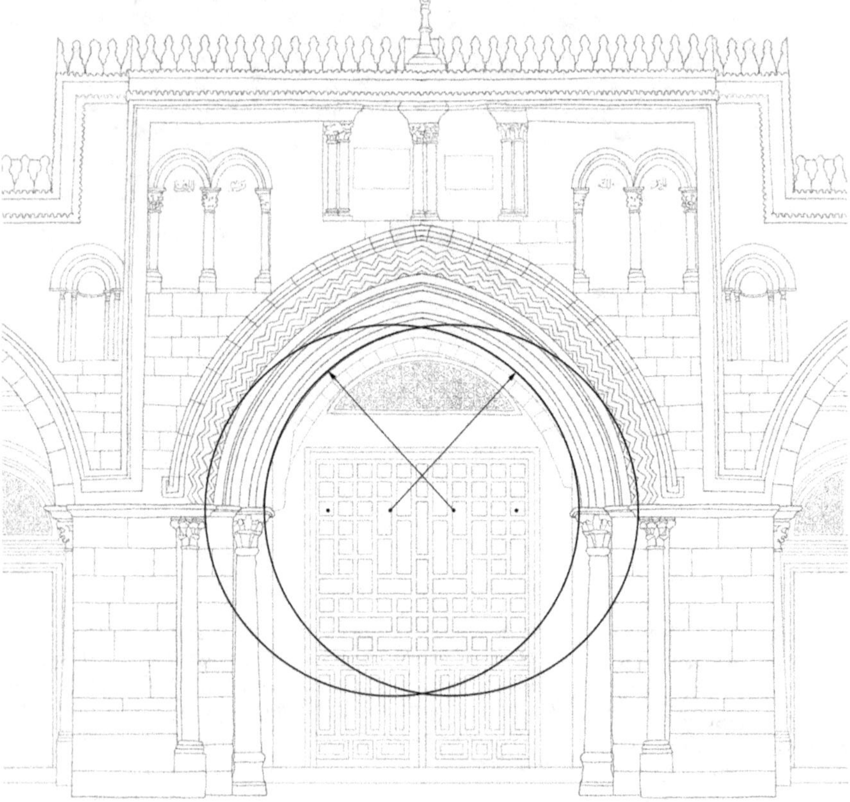

The main openings of the Al-Aqsa Mosque in Jerusalem have pointed arches. You can see in the above drawing that some of the blind niches in the walls have three-centred arches similar in shape to the very much larger Tāq Kasrā on the previous page. But the main archway has a crisper geometry composed of two arcs of circles meeting in a point.

The pointed arch became the defining feature of Christian architecture in the Middle Ages. There is, for example, an arch similar to the Al-Aqsa arch in the fourteenth-century Palazzo Chiaramonte-Steri in Palermo, Sicily (right). From the twelfth century, pointed architecture developed into the great cathedrals of Europe. This architecture later became known as Gothic (see pages 154–61).

The centres of the Al-Aqsa arch's arcs are at the second and third fifths, as are those of Chiaramonte's arch below.

POINTED VAULT AND FLYING BUTTRESS
lines of force

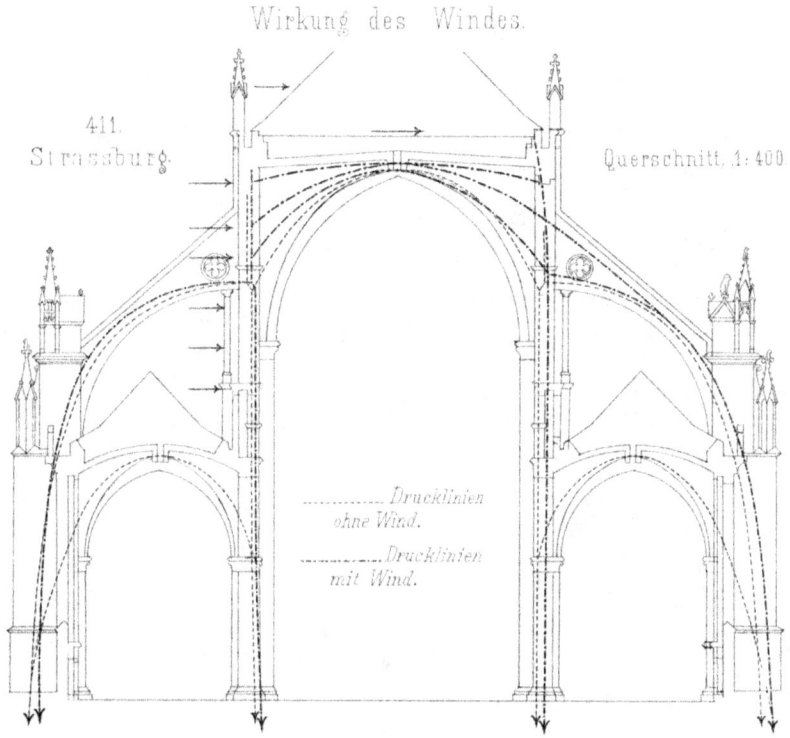

This is a section through the Gothic cathedral at Strasbourg (which was in Germany when this drawing was published but which is now in France). It is from Georg Gottlob Ungewitter's *Lehrbuch Der Gotischen Konstruktionen* (*Manual of Gothic Construction*, 1901). The section, taking into account wind pressure from one side, indicates the lines of force passing from the weight of the vaults and arches down through the columns and flying buttresses to the ground. It can be seen that those lines of force do not follow either the curve of a circular arch nor the arcs of a pointed arch. They follow lines that are curves with radii that vary progressively from one end to the other. These curves resemble parabolas, hyperbolas or catenaries. The differences between these types of curve and their relation to structural geometry are illustrated in the following pages. It can, however, be seen that the Tāq Kasrā vault on page 73 follows the lines of those structural forces better than either the circular or the pointed arch.

This section through Strasbourg Cathedral illustrates how flying buttresses transfer the lateral forces from the high vaults over the central nave along curved paths to the ground.

A catenary curve is 'parametric' in that it is determined by and can be calculated according to the gravitational parameter µ.

CATENARY CURVE
a curve determined by gravity

We are inclined to seek authority for our decisions somewhere other than in our own will. We like to follow seemingly natural rules. By dispelling whim they seem to provide the security of predictability and extrinsic determination. The authority for the round arch is the circle described by a peg and string or pair of compasses (page 67). The authority – the rule by which it is disciplined – of the three-centred arch on pages 72–3 is the pegged and knotted string by which the three arcs of circles are seamlessly drawn. The catenary curve (above) is a curve determined by gravity; it is the line a chain follows when hanging freely from two points. Because gravity determines its geometry the catenary is considered an apt curve for architecture that resists gravity.

The word 'catenary' originates in the Latin word for chain, catena. Its curve is determined by gravity. Inverted, it offers a form that has been appropriated for arches and bridges following the argument that its self-weight loads are evenly distributed and channelled to the ground without distortion.

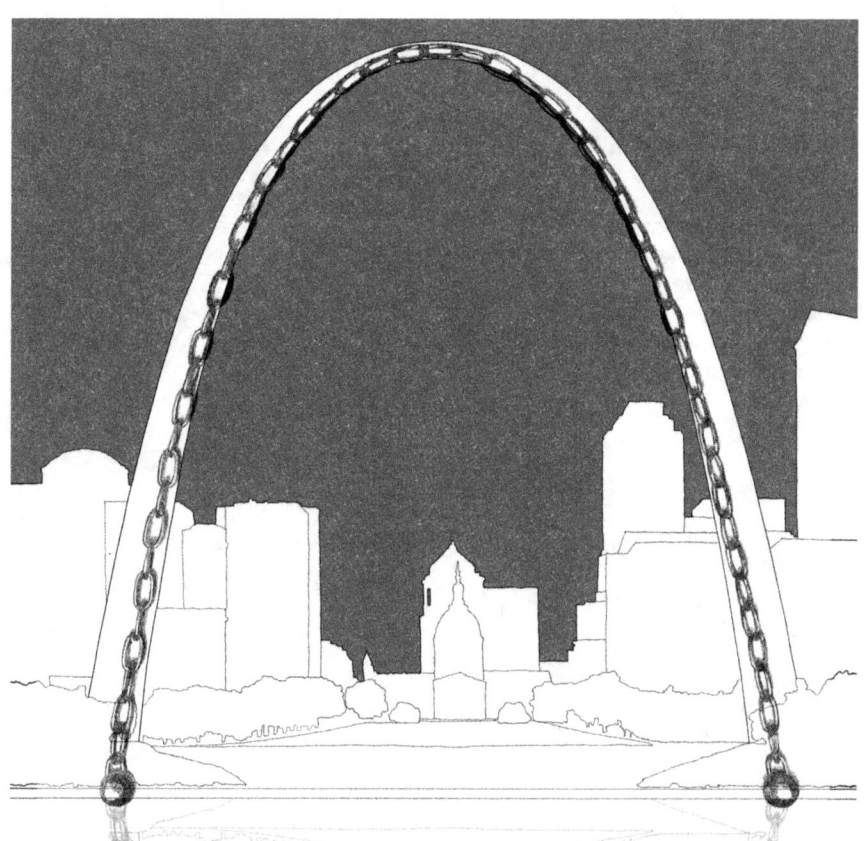

CATENARY CURVE
Gateway Arch, St Louis, Eero Saarinen, 1965

The Gateway Arch in St Louis, which Eero Saarinen designed in 1947, is an example of a slightly amended inverted catenary. It is a free-standing arch 630 feet (192 metres) in both width and height. According to an analysis by Murray Bourne (see the explanation referred to below), the arch's curve is a 'flattened' catenary, which is the curve of a hanging chain that is thinner in the middle than at the ends. There is something poetic, as well as engineeringly pragmatic, about a curve determined by the force it is intended to oppose.

Construction of the Gateway Arch was begun in 1963 and completed in 1965. It stands on the bank of the Mississippi river, which, when calm, reverts the catenary curve to its hanging form by reflection.

You can find a mathematical explanation of how the Gateway Arch is a 'flattened' catenary rather than a parabola at:
www.intmath.com/blog/mathematics/is-the-gateway-arch-a-parabola-4306 (March 2018)

CATENARY CURVE
Casa Milà attics, Barcelona, Antoni Gaudí, 1912

Half a century before Saarinen's Gateway Arch, the Spanish architect Antoni Gaudí had been interested in catenary curves for masonry arches. He liked their shape as well as their structural efficiency.

The attic of the apartment block called Casa Milà (or La Pedrera; 1912; see also page 125) in Barcelona is a warren of catenary arches and vaults built in brick (right).

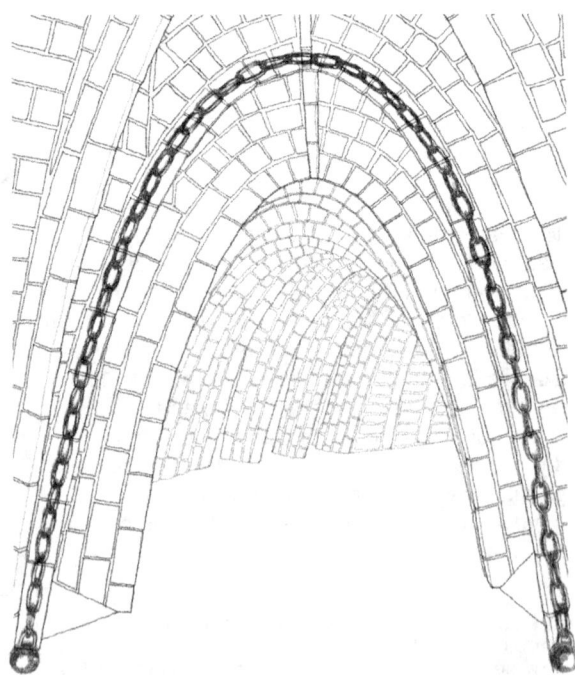

The Casa Milà arches are also examples of curves constructed with straight materials (see pages 95–106). Where the curves become tighter at the top of the arches the bricks are cut into smaller wedges. The increase in the number of mortar joints allows the curve to tighten.

CATENARY CURVE
Gaudí's chain models

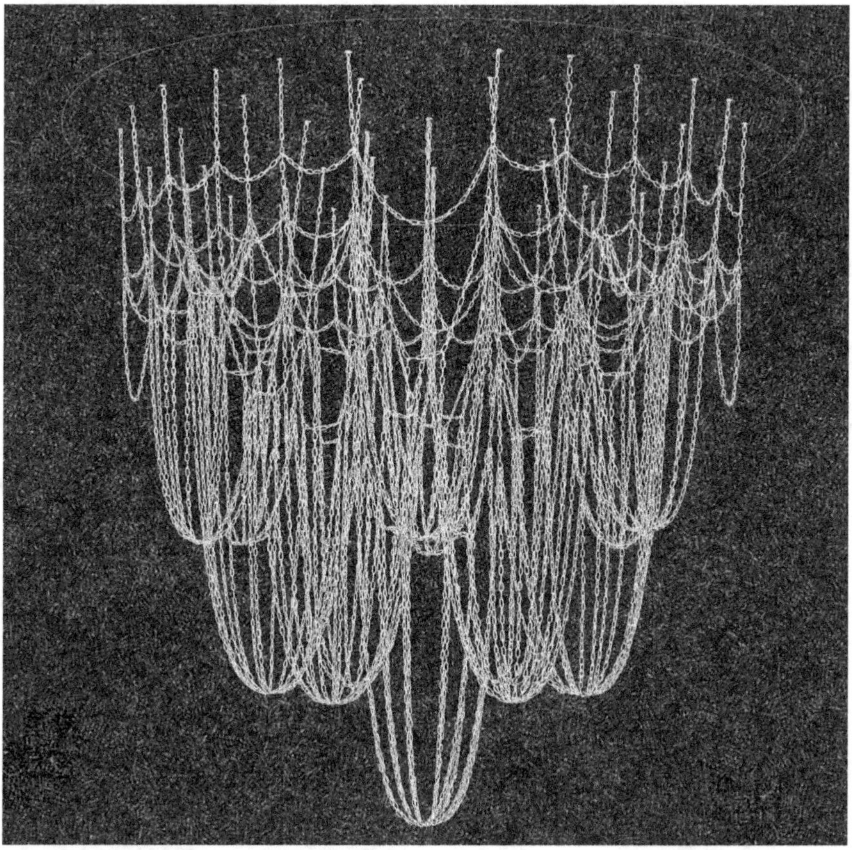

Gaudí explored the possibilities of catenary arches and vaults using chain models. Some of these were very intricate. They formed the bases for his designs for the cathedral of Sagrada Familia in Barcelona (1883–; construction continues) and for the church of the Colònia Güell (1899–1915; only the crypt was completed).

The models – multiple chains hanging according to gravity – provided the forms, when inverted, for the structural curves of the buildings.

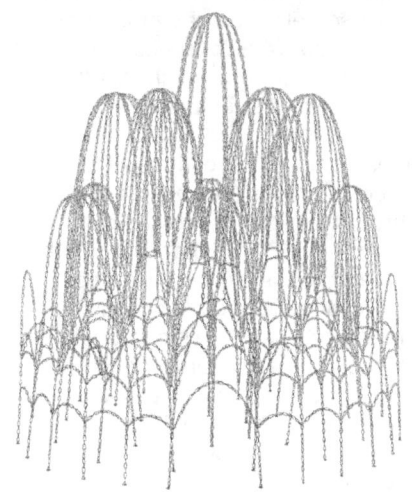

PARABOLIC CURVE
different from a catenary curve

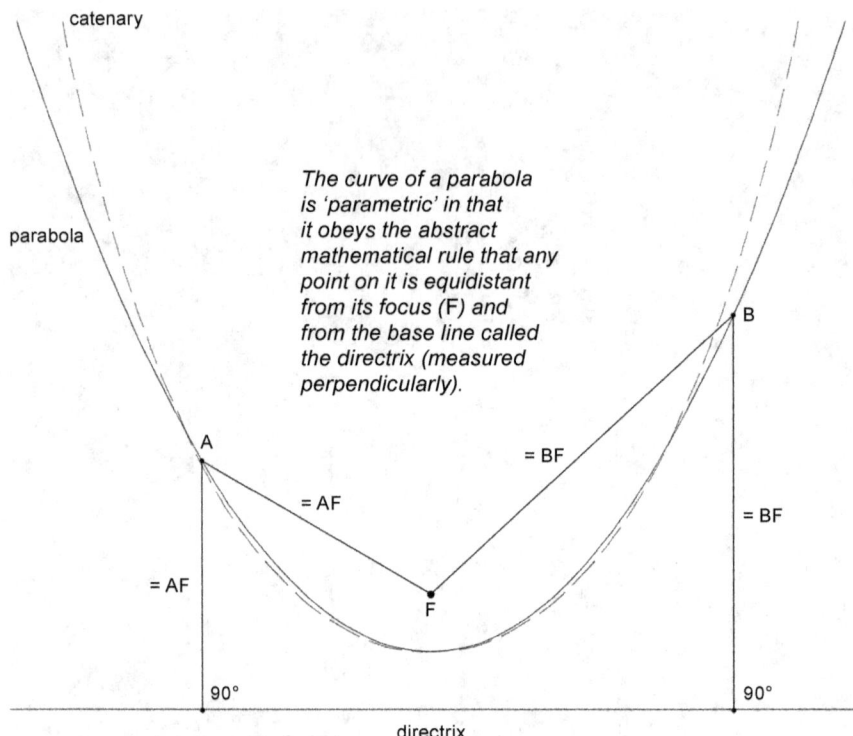

The curve of a parabola is 'parametric' in that it obeys the abstract mathematical rule that any point on it is equidistant from its focus (F) and from the base line called the directrix (measured perpendicularly).

Another curve used in architecture is the parabola. It looks like a catenary curve but is subtly different (see above). It is constructed according to a geometric rule rather than by gravity. But when carrying loads in addition to its own weight it is the most efficient structural curve.

A parabola is one of the so-called 'conic' sections (as shown in the three-dimensional drawing on the right). A section cut parallel with the cone's base is a circle. A section at an angle across the cone is an ellipse. A section cut through the cone's base at any angle is a hyperbola. The parabola is a special case of hyperbola in which the section is cut parallel to the cone's side.

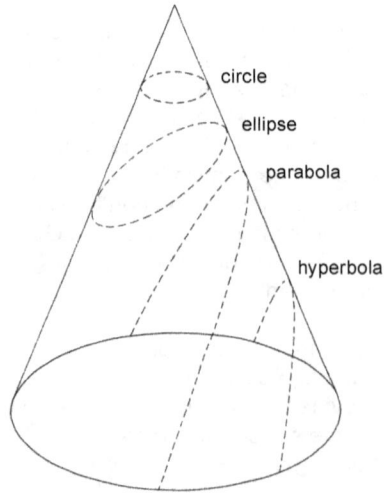

SUSPENSION BRIDGES
when a catenary becomes a parabola

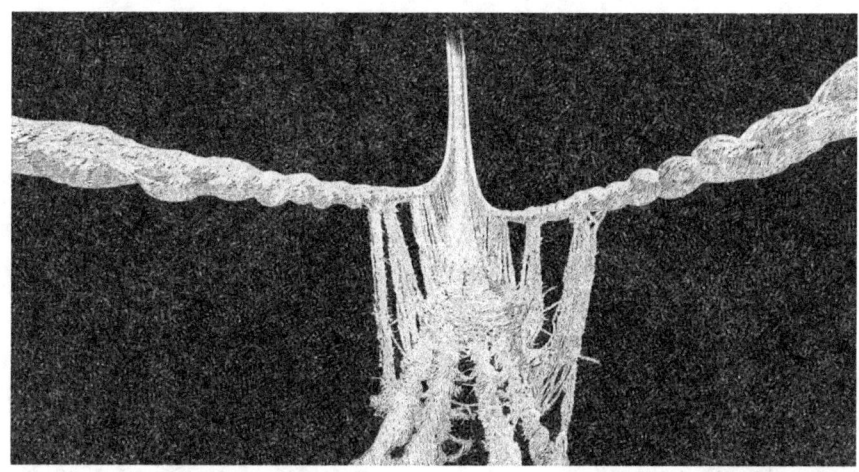

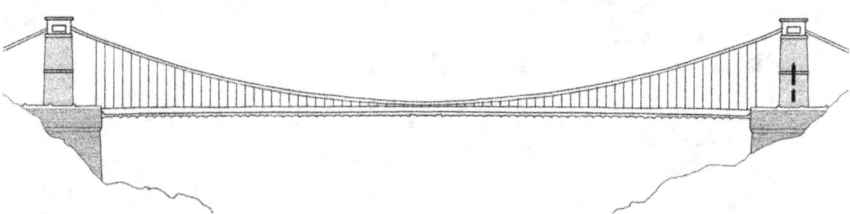

A rope bridge across a chasm (top), so long as no one steps onto it and thus changes its geometry, describes a catenary curve because gravity is acting only on its own weight. When a rope, chain or cable is also supporting a road deck then the curve becomes a parabola.

The chains of Isambard Kingdom Brunel's Clifton Suspension Bridge in Bristol (middle; 1864) describe a parabolic curve. The towers which support the chains at each end of the bridge are also penetrated by stone arches that are not composed of arcs of circles. One of these (the nearer in the drawing on the right) is an inverted catenary while the other appears to be an arch formed of half an ellipse.

PARABOLIC CURVE
two Newcastle bridges

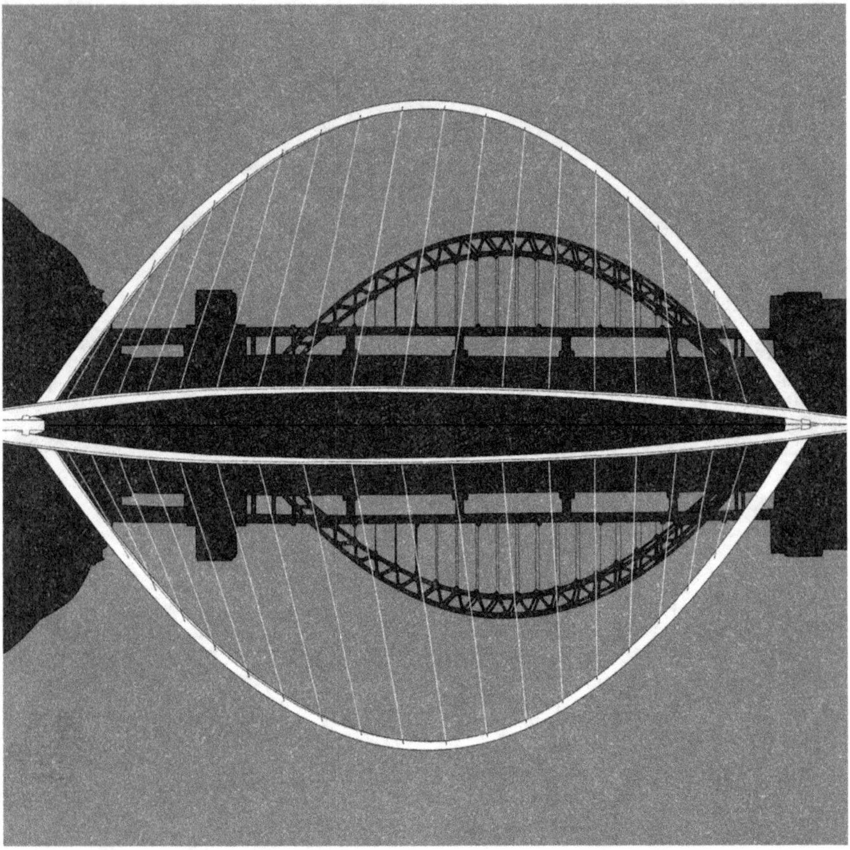

Completed in 2001, the Gateshead Millennium Bridge across the Tyne (above) was designed by WilkinsonEyre architects with Gifford structural engineers. The arch tilts to pivot the curving pedestrian walkway upwards so that ships may pass (right).

The parabolic curve of the Millennium Bridge echoes that of the old Tyne Bridge silhouetted in the background (above). The older bridge was designed by Mott, Hay and Anderson and opened in 1928.

In both cases – the Millennium Bridge as well as the older Tyne Bridge – it seems the parabolic curve was chosen because of its aesthetic beauty as well as practical structural efficiency.

The Millennium Bridge echoes the older bridge.

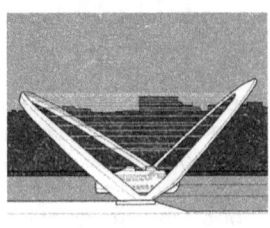

The parabolic arch of the Millennium Bridge tilts to lift the curved walkway so that boats may proceed along the river.

HYPERBOLIC PARABOLOID
parabolas in three dimensions (see also page 100)

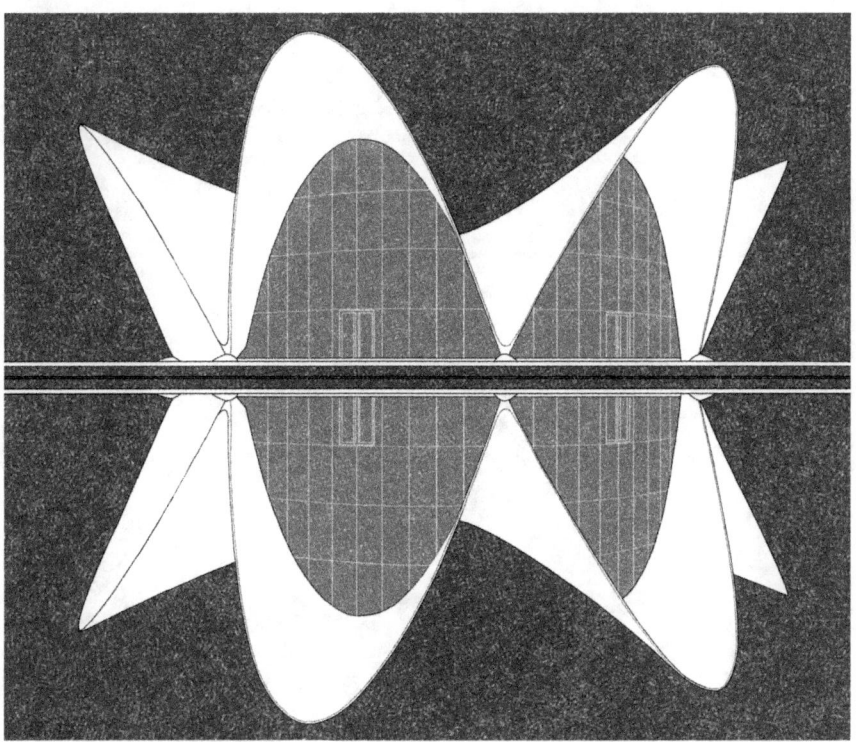

Above is one of the buildings at the City of Arts and Sciences in Valencia, Spain, designed by Santiago Calatrava and built between 2003 and 2010. Its steel-reinforced concrete shell roof is a composition of hyperbolic paraboloids.

Calatrava was inspired by the designs of the Spanish-Mexican engineer Félix Candela. Candela's roof for Los Manantiales – a restaurant in Mexico City (1958, right) – was a radially symmetrical array of vaults derived from hyperbolic paraboloids, constructed as a thin concrete shell. (See also page 101.)

CONCRETE SHELL
Rubber Factory, Wales, ACP and Ove Arup, 1951

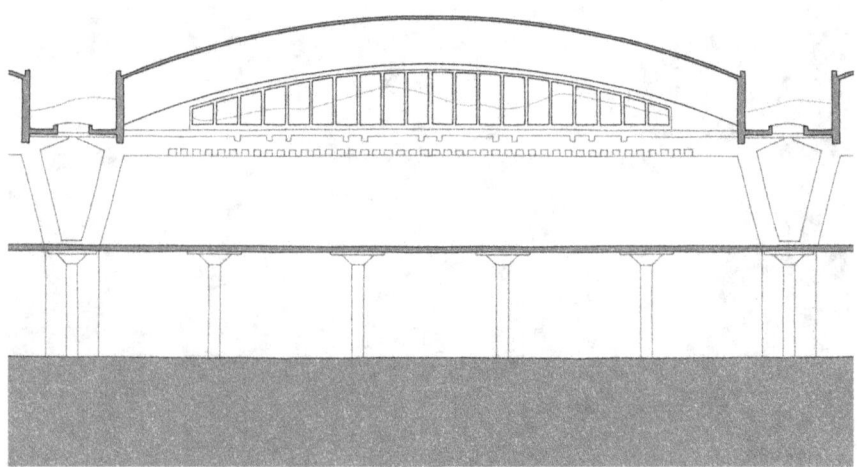

section

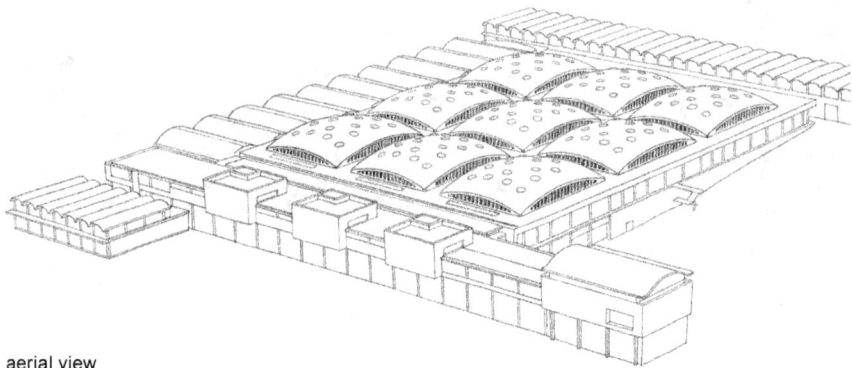

aerial view

The Brynmawr Rubber Factory (above) was designed after the Second World War by the Architects Co-Partnership with engineer Ove Arup. It was demolished in 2001 despite having been previously listed as a building of historic importance. Its main space had nine domes, the concrete shells of which were less than 100mm (4 inches) thick even though the longer span was nearly 26 metres. The whole building was something of an exercise in curved concrete roofs. The boiler house (right), which has survived, has an elliptical concrete roof.

boiler house

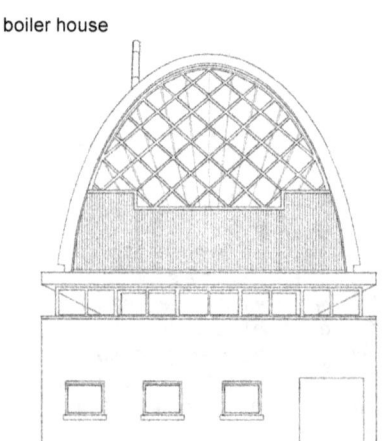

INFLATION CURVES
Sports Dome, Malvern, Michael Godwin, 1977

Inflated things such as balloons find it very difficult not to be curved. Air pressure stretches the fabric of balloons into curves. Some architects and engineers have experimented with using strong inflated balloons as formwork for concrete domes. Some have even found ways of inflating concrete domes, complete with reinforcement, before the concrete has set. In the 1960s the Italian architect Dante Bini developed a system for inflating concrete domes. Between 1970 and 1990 over 1500 'Binishell' roofs were created. Their diameters ranged between 12 and 36 metres.

Inflation creates curves, as in balloons. Inflated balloons have their own structural stability. Air-sealed fabric tents can be inflated like balloons, without need for additional support. Alternatively, simple buildings can be constructed with walls of sealed pockets, which, when inflated, give the building its shape (below right).

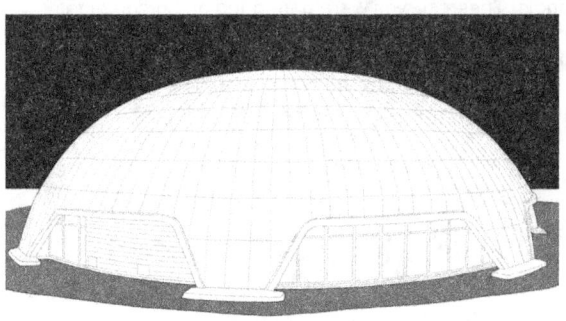

Edinburgh Sports Dome, Malvern, Worcestershire

The Edinburgh Sports Dome in Malvern, England was designed by Michael Godwin and built in 1977. It used ideas of inflated structures inspired by the work of Dante Bini. Concrete was laid over reinforcement and then inflated into a dome whilst still wet.

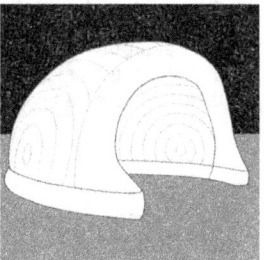

The 'Luna Pod' (above) is made by a British firm called Evolution Dome. Indicating the structural possibilities of inflatables, they developed a system they call 'Air Beam' technology, which is used in a range of structures.

TWO LONDON OLYMPIC VENUES
with very different curved structures

Here are two of the arenas constructed for the 2012 London Olympic Games. The Aquatic Centre (below) was designed by Zaha Hadid Architects; the Velodrome (opposite) by Hopkins Architects. Both have double-curved ('happy–sad'; see page 119) roofs. Glancing at the buildings you could think that their structures might be similar. But they are radically different. Their difference illustrates divergence in theoretical approaches to producing curved architectural structures.

Appropriately, the long section of the Aquatic Centre roof has the profile of a swimming seal.

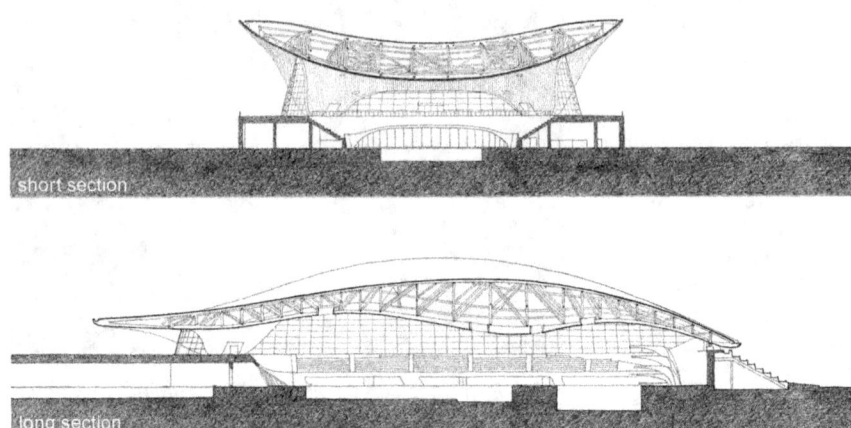

short section

long section

The form of the Aquatic Centre is the more complex of the two. Its lower curves (the swimming hall's ceiling) vary more than the upper, giving the roof's long section a profile a little like a seal swimming (above). The varying ceiling heights relate to the swimming and the diving pools. These curves were generated on 3d computer software by eye and mainly for aesthetic and expressive effect. To achieve this complex sinuous form, the roof's structure (below) is a complex lattice of steel beams, each of which is straight. This lattice is encased by the cladding and ceiling. The resulting roof is a heavy volume resting on point supports at each end (like a dolmen).

The hidden structure is a complex lattice of straight steel beams dimensioned by computer to support the roof's sinuously curved outer form (right).

In this double-curved roof, the design is led by aesthetics. The structural system follows, and has to cope with what the aesthetic design demands.

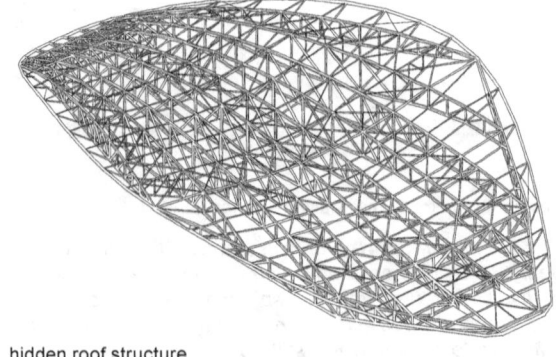

hidden roof structure

The Velodrome roof might appear similar in its double curvature but structurally it is very different. By comparison with the roof of the Aquatic Centre, it is thin and its curves are regular, swelling up to the centre over the track's long section, and sweeping upwards to the edges in the short section (below).

The structure is not a lattice of structural beams, but a mesh of two sets of tension wires stretched across the arena at right angles (bottom). This mesh supports the roof cladding. The wires are attached to a ring beam that determines the geometry of the roof's edge. This ring beam absorbs the tension of all the wires. The curved geometry of the roof is a result of the generally downward pull of the long section wires countering the generally upward pull of the short section wires.

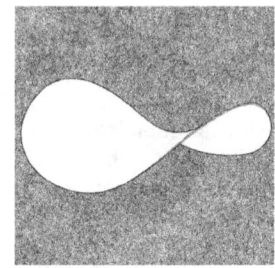

The Velodrome's roof is similar in form to a hyperbolic paraboloid potato crisp.

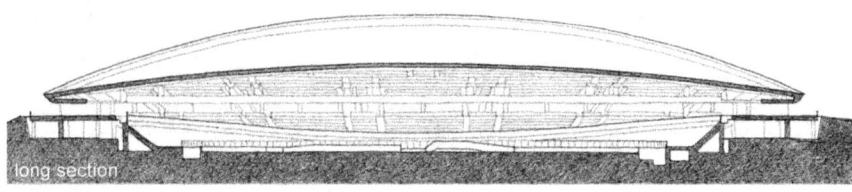

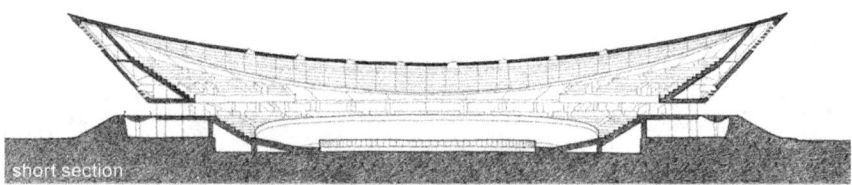

The structural philosophy of the Velodrome might be described as the opposite of that of the Aquatic Centre roof. Here the aesthetic is the result of a structural system rather than its driver. The form is the desired consequence of that system's internal forces interacting with gravity. As such its curves are innately structural; they are 'structural curves' similar in their generation to arches though deriving from tension rather than compression. By contrast, the curves of the Aquatic Centre roof are not structural in origin; they are 'constructed curves'; in which aesthetic dominates form.

The structure of the Velodrome roof is a mesh of two sets of tension wires (left). One set is pulling the curve downwards in opposition to the other, which is pulling upwards. The hyperbolic paraboloid form of the roof is a result of the interaction. Here, it is the structural forces themselves that contribute to the form of the curving roof.

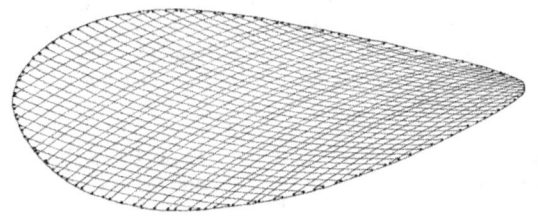

mesh of tension wires

CURVE

BENDING
reeds and rods

Curves result from bending a flexible material. In regions where suitable materials are available, traditional structures were developed with curved forms, such as these houses in Iraq (right). The same principle is used in modern tents (below).

The cavity of a modern tent is created by metal rods held bent in narrow pockets in the tent's fabric.

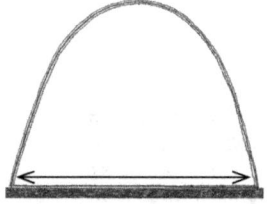

The curve of the reed bundles of the traditional Iraqi houses (right) is maintained by their being bedded in the ground. The curve of the tent's metal rods (above) is maintained by tension in the ground sheet, the tailoring of the tent and tension in the stabilising guy strings.

The above images show the construction process of the structure of a traditional type of house built along the Tigris and Euphrates rivers in south Iraq. The structure is formed of bundles of the giant reeds that grow along the banks. These are bedded quite deeply into the ground and then bent to form roughly parabolic curves bound together at the apogee. These principal structural arches are then covered with mats woven from split reeds to form the roof.

Bent structures, like cruck frames (pages 64–5), have their own innate structural stability and resistance to sideways forces such as wind.

See Bernard Rudofsky – *Architecture Without Architects*, 1964.

BENT STRUCTURE
traditional shelters and gridshells

Some of the simple shelters of traditional cultures around the world are made by bending saplings, planting their ends in the ground and then tying the hoops together where they cross (right). Wigwams, teepees, lodges in North America, Aborigine humpies in Australia, can be made in this way.

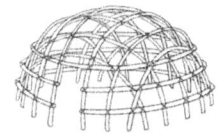 

Many traditional shelters around the world exploit the structural strength of tree saplings bent into arches and formed into lattices. The North American native wigwam or lodge (above) is such a structure. It would be covered with bark or hides.

The walls of yurts and gers in middle Asia – nomadic shelters – are framed of expandable lattices curved into a circle (right). The lattices can be folded for travelling.

The Weald and Downland Open Air Museum in England has a large shelter designed by Cullinan Studio and built in 2002. It has a gridshell structure (below). Oak laths were made into a lattice and then bent into a vault.

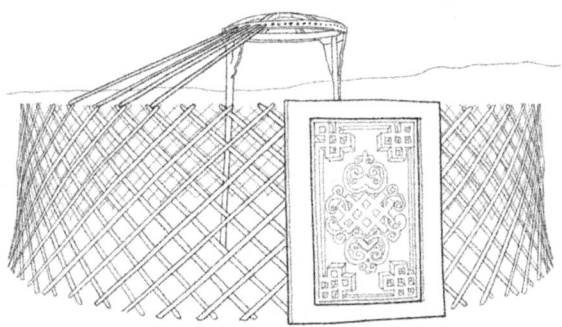

Yurts have wall lattices that can be folded for travelling.

The gridshell of the Weald and Downland Museum building was made flat on scaffolding; then gradually bent down to meet the ground and secured in place.

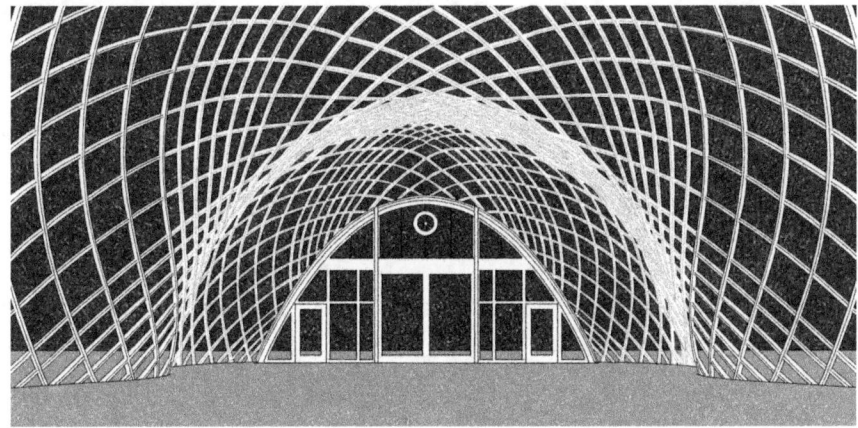

BENT CONSTRUCTION
Artist's Studio, Somerset, David Lea, 1989

In the 1980s David Lea designed a tiny studio using bent saplings to provide a framework for the walls and support to the thatched roof.

The saplings were bent and thrust into the ground. A lattice of smaller saplings was then woven over the structure and plastered inside and out.

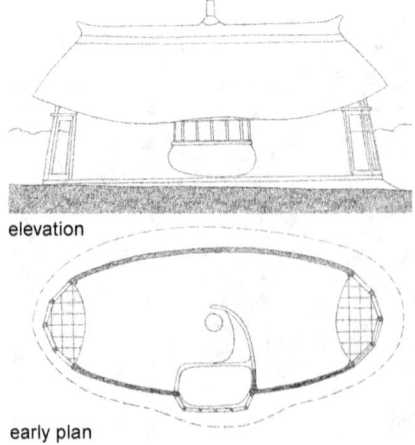

elevation

early plan

section

MONOCOQUE
structural skins

The structure and construction of many buildings is analogous to that of the human body. There is a skeleton that provides the structure, over which there is a weather-proof skin. The structure can stand independent of the skin. By contrast, in monocoque structures either the skin *is* the structure or is integral to it – maybe a crustacean's shell is a good analogy. Monocoque structures were developed in boat building and in aircraft design where lightness is desired. In both, curves are essential to their hydrodynamic or aerodynamic performance. But curves are also essential for structural strength (as in an egg's shell).

Some architects have explored the potential of monocoque structures in architecture. For example, when Future Systems were asked to design a new store for Comme des Garçons in New York they made the entrance as a tunnel (below). The tunnel was formed of aluminium sheets beaten into a curved monocoque structure. Its own curves give it the necessary structural strength, stiffened only by slight external ribs.

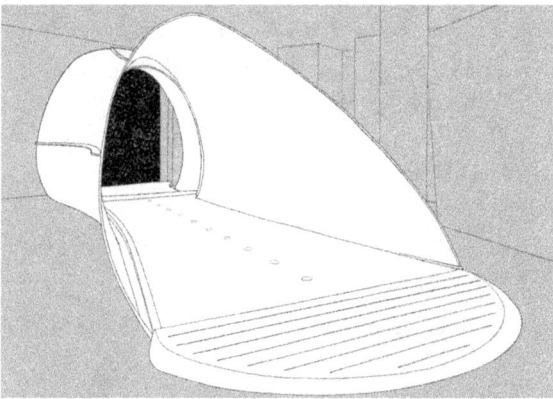

Comme des Garçons store, New York, Future Systems, 1998

Also in the 1990s the same architects designed a Media Centre for Lord's Cricket Ground in London (below). Its aluminium sections were prefabricated in a boat yard. It is a semi-monocoque structure: the skin is structural, but its strength and stability depend also on an integrated structural frame.

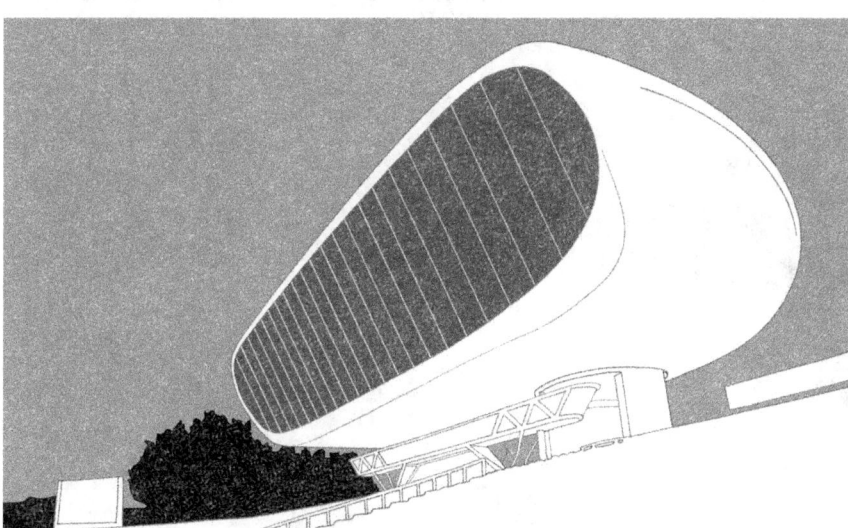

Lord's Media Centre, London, Future Systems, 1999

3D-PRINTED CURVES
building like termites

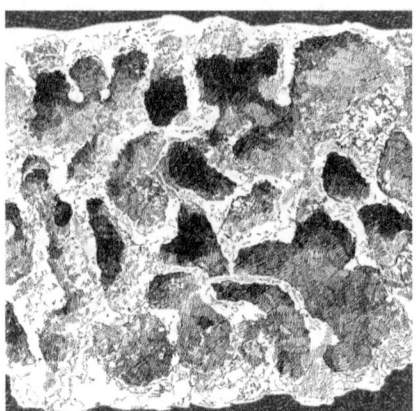

Termites make their mounds (above) and their labyrinthine passageways (above right) by building up layers of moist earth that then set like concrete. This means that the geometry of their architecture does not need to obey the rectangular geometry of making but grows in curves.

Some types of 3d printing, especially those used in architectural model-making, and even in the making of full-sized buildings, operate on a similar method. Material is laid down, thin layer by thin layer, controlled by computer.

Various methods and machines for 3d printing buildings are under development as I write this Notebook. One example is the Curve Appeal Home by WATG (below), designed in 2016 but programmed to begin construction in 2018.

Human architects can be envious of the capacity of insects and other creatures to create free-curving structures.

3d printing goes some way to making such free-curving structures possible.

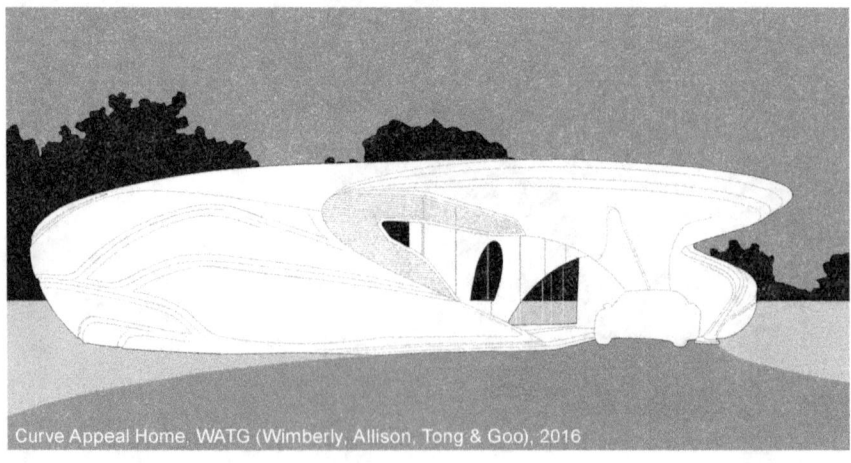

Curve Appeal Home, WATG (Wimberly, Allison, Tong & Goo), 2016

STRETCHED FABRIC CURVES
tents and awnings... and chewing gum

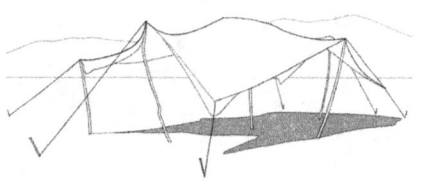

fabric shade shelter

fabric desert tent

Polish Pavilion, Izmir International Fair, Oskar Nikolai Hansen, 1955

There have probably been tents since prehistoric times. Supported by poles and pulled taut by guy ropes, normal non-stretch fabrics – hide or woven cloth – form curves. These can be exaggerated and developed into dramatic sculptural form when stretchable fabric is used (below).

Hansen's Polish Pavilion (above) appears to be composed of hyperbolic paraboloids (see pages 100–101) but actually comprises ranges of twisted striped fabric tents.

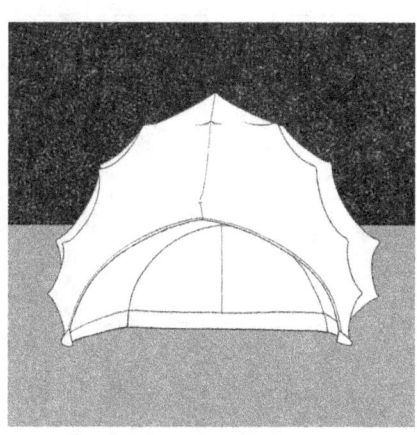

This is a simple event pavilion mass-produced by the multinational firm Stretch Marquees and Fabric Structures. It is called a 'stretch pod'.

Complex three-dimensional stretched chewing-gum-like forms can be made. This design is by the British firm of designers and architects, AirSculpt.

STRETCHED FABRIC CURVES
Sackler Gallery, London, Zaha Hadid, 2009

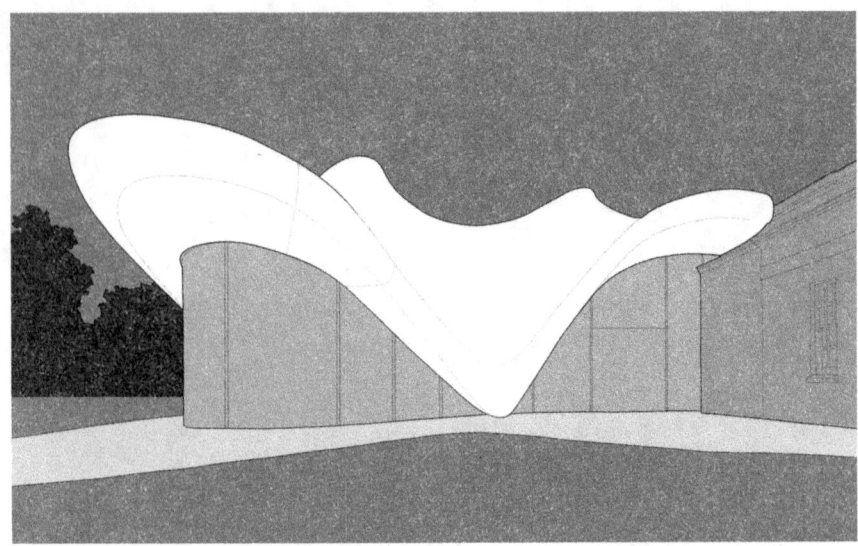

The Sackler extension to the Serpentine Gallery in London's Hyde Park was designed by Zaha Hadid and built in 2009. Its sculptural curves are formed of fabric stretched over a sinuous armature of metal tubes. One layer of fabric forms the external surface (above), another the internal (below). Light enters through rooflights incorporated into the sculptural equivalents of the tent poles.

Stretching fabric over a metal armature bent into curves is an effective way of creating complex three-dimensional surfaces. It is also more economical than trying to fabricate complex curves from stiff non-stretch materials.

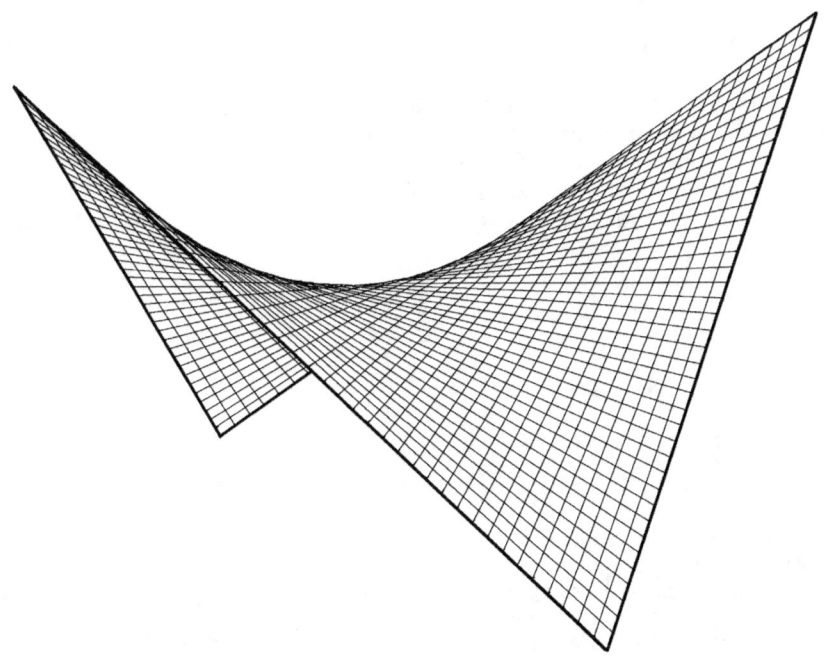

CURVES FROM STRAIGHT LINES

Given there is a predisposition toward straight lines and right angles in the world which we create for ourselves, and that this predisposition clashes with an attraction to curves, it is understandable that those who wish to make curves have experimented with ways of resolving that clash by generating curves from straight lines. Some of these experiments have involved arranging relatively small elements, such as bricks or tiles, so that they form as tight a curve as their rectangular geometry allows. But other experiments derive from geometric constructions of straight lines such as the hyperbolic paraboloid illustrated above, in which an array of straight lines in two directions forms an apparently curved surface. The straight lines of such diagrams can be converted into straight structural elements – timber rafters or steel beams – to form curved roofs.

CONICAL ROOFS
roofing a roundhouse

Attempts to create curved surfaces with straight, or relatively straight, pieces of building material stretch back into prehistory. The easiest way to create a building that is curved is to build a cone, as in a forest bivouac or a teepee.

 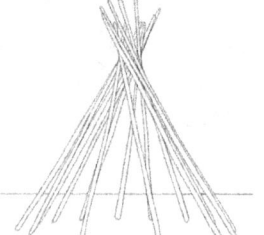 

The teepee structure is a way of roofing curved (circular) plans with straight elements. It is a strategy used in prehistoric roundhouses (see page 63) but it can also be applied in large sophisticated steel and concrete modern structures, as in the case of the Liverpool Metropolitan Cathedral (right).

More inventive developments of the conical structural strategy can be found in spiral or 'reciprocal' structures (below). These can be simple or multi-layered. Conically pitched, tangentially arranged, and seemingly defying gravity, the rafters of these structures bear on each other without the need for a central post.

Liverpool Metropolitan Cathedral, Frederick Gibberd, 1967

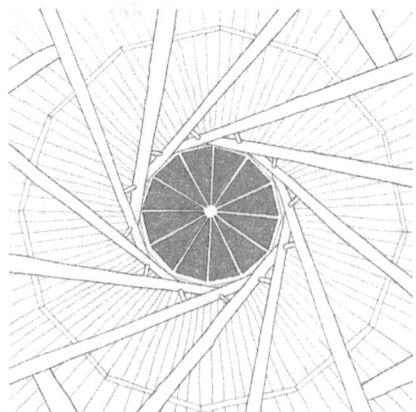
reciprocal roof, Hill Holt Wood, Lincoln School of Architecture (Dr Behzad Sodagar) 2006

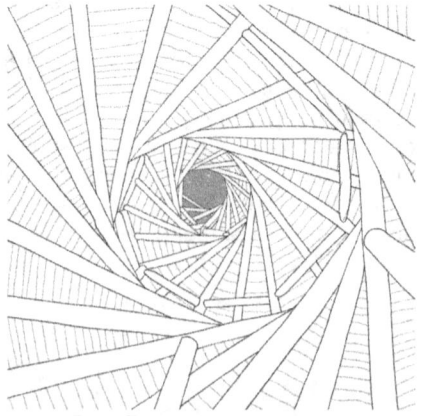
house, Fri og Fro ecoVillage, Egebjerg, Denmark, Poula-Line Schmidt, 2012

VAULTING TUBES
using straight components for curve structures

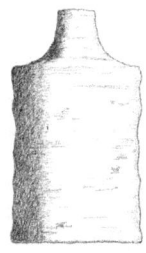

For over two thousand years, builders in north Africa have made roof structures using fired clay tubes (above). These slot one into the last and may be formed into vaults of varying radii (right). The Roman Empire borrowed this technology, using it in forts along Hadrian's Wall. The tubes can even make groin vaults, as in this Roman villa in Tunisia (top right).

In India, similar structures are made using guna tubes (below). They too are clay and slot together to form vaulted roof structures (right).

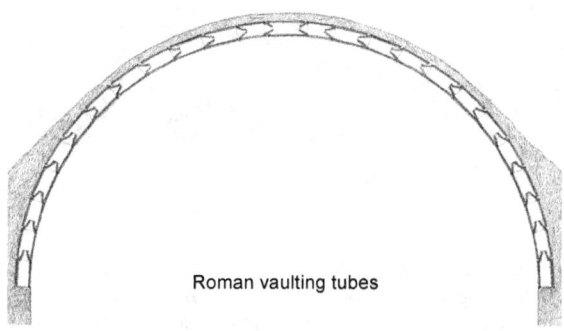

Roman vaulting tubes

Roman vaulting tubes (above) and Indian guna tubes (below) are straight components but, because of the way they fit together, they can be used to build curved vaulted roofs of varying radii and therefore cover different spans between supporting walls.

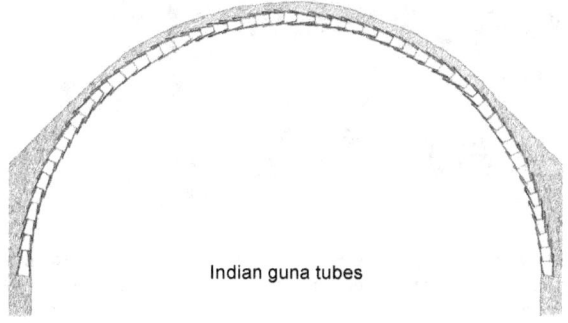

Indian guna tubes

CURVE

MAKING CURVES WITH BRICKS
Brick answers Kahn's question (asked on page 12)

'And Brick says to you, "I like an Arch." And if you say to Brick, "Look, arches are expensive, and I can use a concrete lintel over you. What do you think of that, Brick?" Brick says, "I like an Arch." And it's important, you see, that you honor the material that you use…'

Louis Kahn, to students, 1971.

If pressed to elaborate, to express a more detailed opinion of his innate likes and dislikes, Brick might continue:

'But since you ask… I am not sure that in my standard form I am best suited to an Arch. I am intractably inflexible… I don't really do curves… I can just about manage gentle ones… and so an Arch is not impossible. It would be better, though, if I was specially shaped to fit the curves needed.'
(continued at the bottom of the page)

Because of the joints, it is possible to lay bricks to gentle curves (below). The more joints the tighter the curve possible.

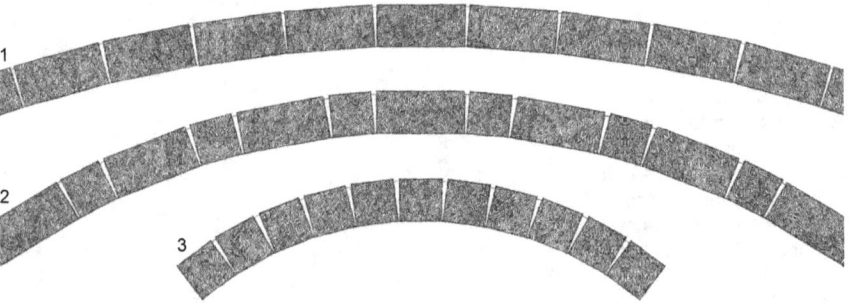

Using stretcher bond (1) a radius of 3.5 metres might be achievable; in Flemish (2) maybe 2.00 metres; and in half bricks (3) 1.00 metre radius. But all are faceted rather than smoothly curved.

Wedge-shaped and curved bricks are less flexible. They are limited to one specific radius (below left), and they cannot be built straight. In a complex curved wall (below) every single brick might be different. Standard rectangular bricks are the most versatile and economical.

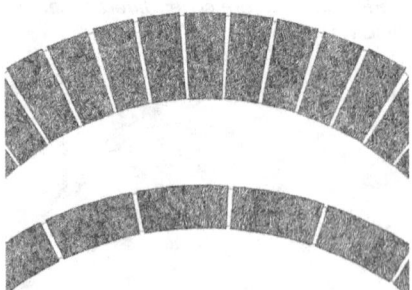

'To make tight curves it might be better to make me wedge shaped or curved!'

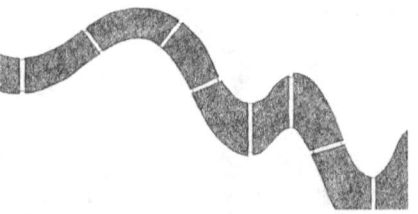

'And for complex curves you would have to make me in lots of special shapes.'

98 ANALYSING ARCHITECTURE NOTEBOOKS

CURVES AND THE BRICK
but you cannot bend a brick

As hinted at on the previous pages, the decision to use an arch may not necessarily be driven by the 'likes' of Brick but by the architect's predilection for the curved expression of dynamic force evident in the curved form of an arch but not a straight lintel.

One of Kahn's great brick buildings is the Indian Institute of Management in Ahmedabad (right). There he used a distinctive combination of both lintel and curved brick arch.

Indian Institute of Management, Ahmedabad, Louis Kahn, 1974

Incurvo House, Oxfordshire, England, Adrian James, 2016

Adrian James Architects managed to produce this curvaceous plan (above) using mainly standard rectangular bricks. The curves are only in plan. The walls are vertical; doors and windows rectangular. .

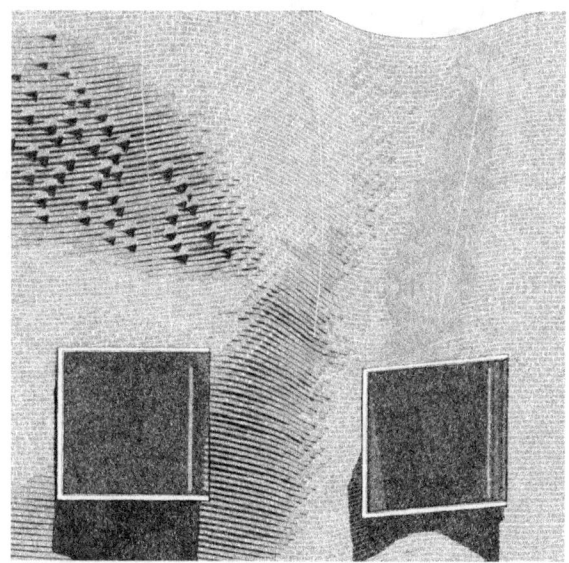

Dr Chau Chak Wing Building, Sydney, Frank Gehry, 2015

Whereas the three-dimensionally curved brick cladding of this Frank Gehry building (above) requires many different special bricks and a complex constructional system. The result challenges whatever likes Kahn's rectangular Brick might have had.

CURVE

CURVES FROM STRAIGHT LINES
geometric constructions

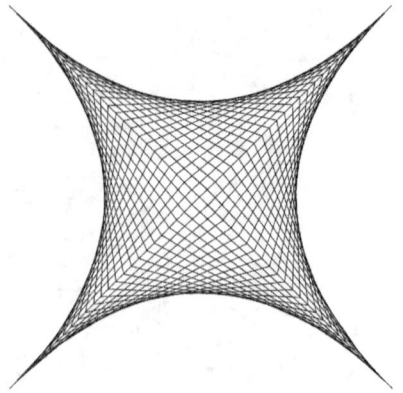

Graphically, it is possible to generate curves using straight lines constructed according to preordained rules – algorithms. All the images on this page are constructed according to the rules stated. But there are infinite possibilities. Because they are constructed of straight lines, translation to architectural construction is relatively easy to understand (if not always achieve). Computer design packages can be coded to apply such algorithms. Computer generated curves are calculated in relation to matrices of points connected by straight lines (see page 106).

above: divide diagonals equally; progress linking lines step-by-step in opposite directions

below: divide diameter equally; progress chords around, 1 space one end, 2 the other

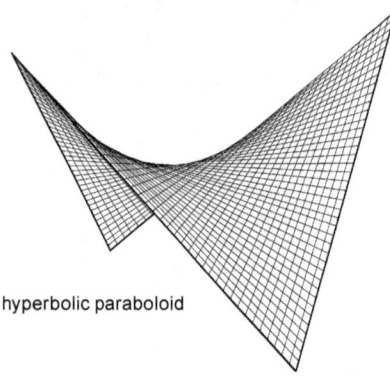

hyperbolic paraboloid

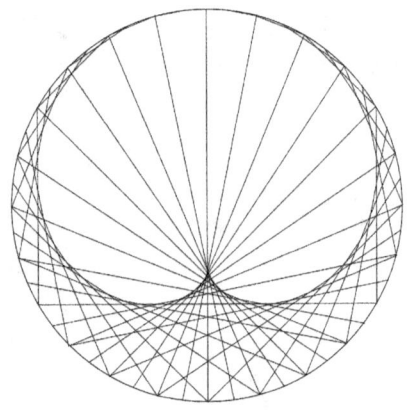

above: repeat in three dimensions

below: twist a cylinder 90° both ways

below: nest squares; reducing 5%; rotating 3°

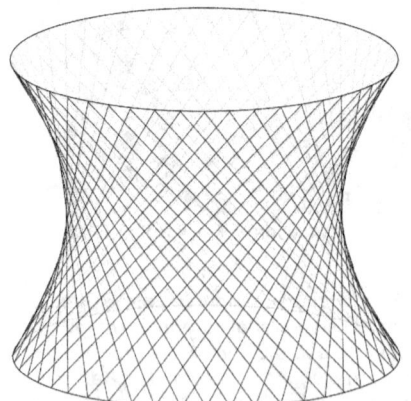

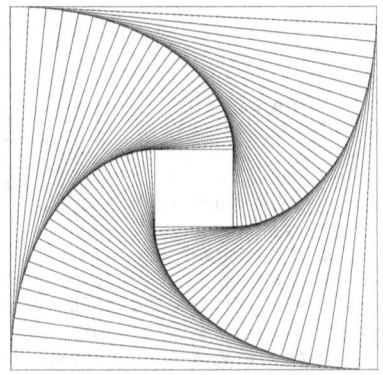

CURVES FROM STRAIGHT LINES
hyperbolic paraboloid structures

hyperbolic paraboloid structure

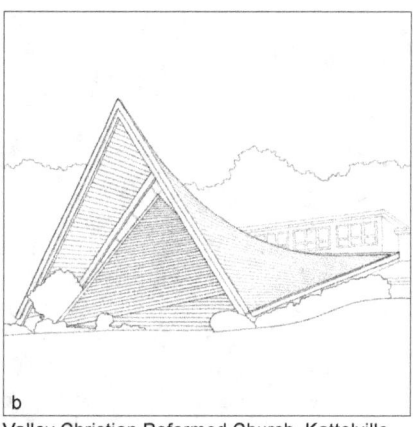

Valley Christian Reformed Church, Kattelville, NY, James Mowry, 1968

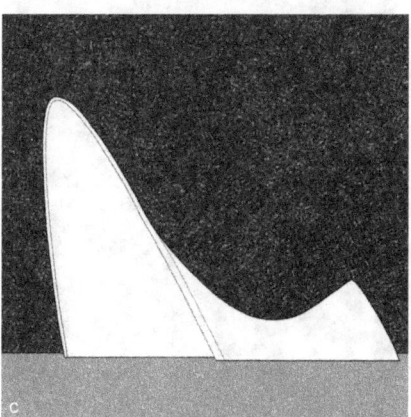

Chapel Lomas de Cuernavaca, Mexico, Félix Candela, 1958

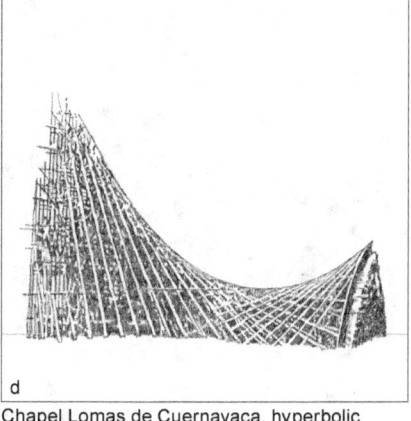

Chapel Lomas de Cuernavaca, hyperbolic paraboloid formwork

Hyperbolic paraboloid structures can be constructed with straight elements... though they may have to be twisted a little (a and b).

This means the formwork for hyperbolic paraboloid concrete shell roofs can be constructed with straight elements (c and d, middle).

Sculptural structures can be derived from the geometric constructions opposite too; such as this tower in Slovenia (right).

right: Vinarium Tower, Lendava, Slovenia, Oskar Virag and Iztok Rajšter, 2015

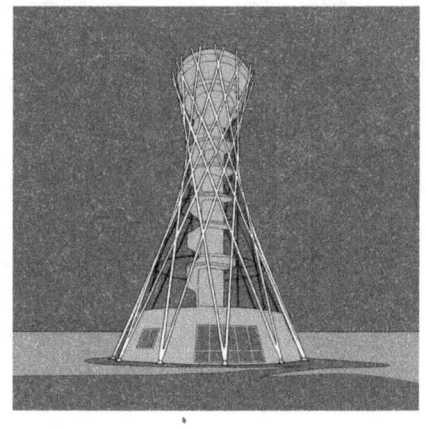

CURVES FROM STRAIGHT LINES
Biosphere, Montreal, Buckminster Fuller, 1967

In the 1940s, 50s and 60s the American architect and engineer Richard Buckminster Fuller popularised what he called the 'geodesic' dome. By creating spherical matrices of triangles he developed strong but light structures ranging in size from the small to the very large. The triangular matrix meant that these apparently curved structures were made completely with straight elements holding flat panes of glass or other material.

Buckminster Fuller designed the United States pavilion (above) for the 1967 World Exposition held in Montreal, Canada. His design was called the Biosphere. It is seventy-six metres in diameter and easily accommodates a seven storey exhibition building. Fuller also developed many smaller versions, including the lightweight transportable structure below.

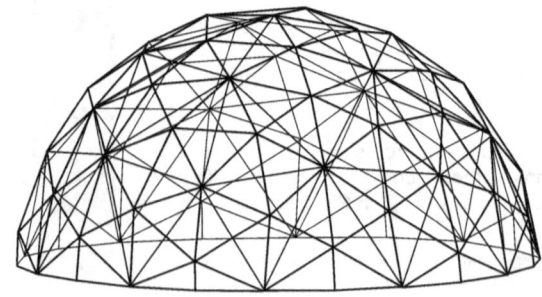

ANALYSING ARCHITECTURE NOTEBOOKS

CURVES FROM STRAIGHT LINES
roof, British Museum, Foster + Partners, 2000

The same principle – of composing triangles to created a faceted, but apparently curved, structure – was used by Foster + Partners when asked to design a roof for the Great Court of the British Museum in London. But whereas Buckminster Fuller's designs followed the regular geometry of a sphere, Fosters had to fit their roof over a space between the edges of a rectangular courtyard and the almost-but-not-quite-central rotunda of the museum's old library. Reconciling these conflicting geometries produced a structure of apparent curves but which, as in the geodesic dome, is composed completely of straight elements and flat panes.

Foster + Partners' roof over the Great Court of the British Museum seems composed of curves and appears to 'balloon' above the parapets as if inflated from below. But its complex engineering – an integrated matrix of triangles modified to fit the subtle irregularities of the existing courtyard – is achieved with straight elements and flat panes of glass.

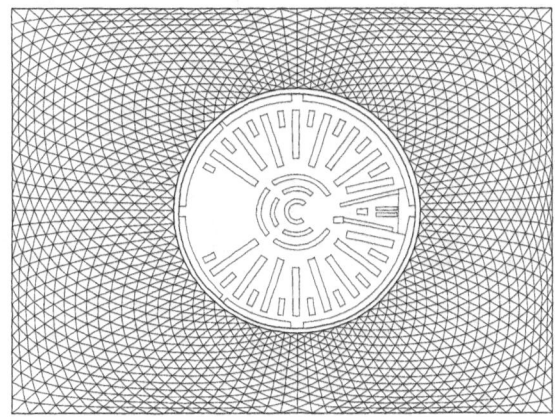

This plan of the roof shows the extent of the courtyard. The galleries of the museum wrap around this space.

CURVES FROM STRAIGHT LINES
sculpture and metaphor

The intricate forms that can be created with algorithmic arrangements of straight elements are engaging because of the surprising effects that can be produced using simple geometric means. One does not expect beautiful curves to be generated by compositions of straight lines. Such arrangements mean that intricate sculptural forms can be created using simple materials, even though the technology of their construction might be complex.

The Lake Garden sculpture illustrates the eye-catching potential of intricate arrangements of straight steel bars to produce curves that change with different viewpoints.

The two examples illustrated here are:

above right:
an urban sculpture in the Lake Garden of Suzhou, China, made by the Beijing Sino Sculpture Landscape Engineering Company, which evokes the image of a windmill or of a swirling flamenco dress.

below right:
the Bridge of Aspiration, designed by WilkinsonEyre in 2004, links the Royal Ballet School with the Royal Opera House in Covent Garden, London.

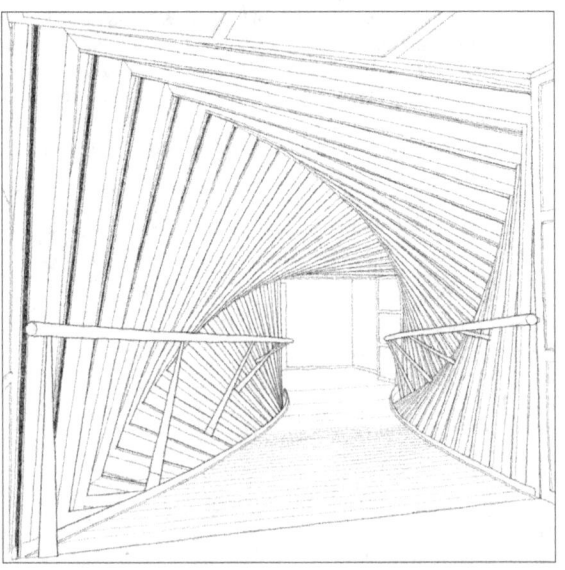

The Bridge of Aspiration is a metaphorical allusion to the pirouettes and graceful movements of ballet dancers. It is composed of twenty three square frames progressively rotated along the length of the bridge.

CURVES FROM STRAIGHT LINES
Organic Cube, Copenhagen, Søren Korsgaard, 2009

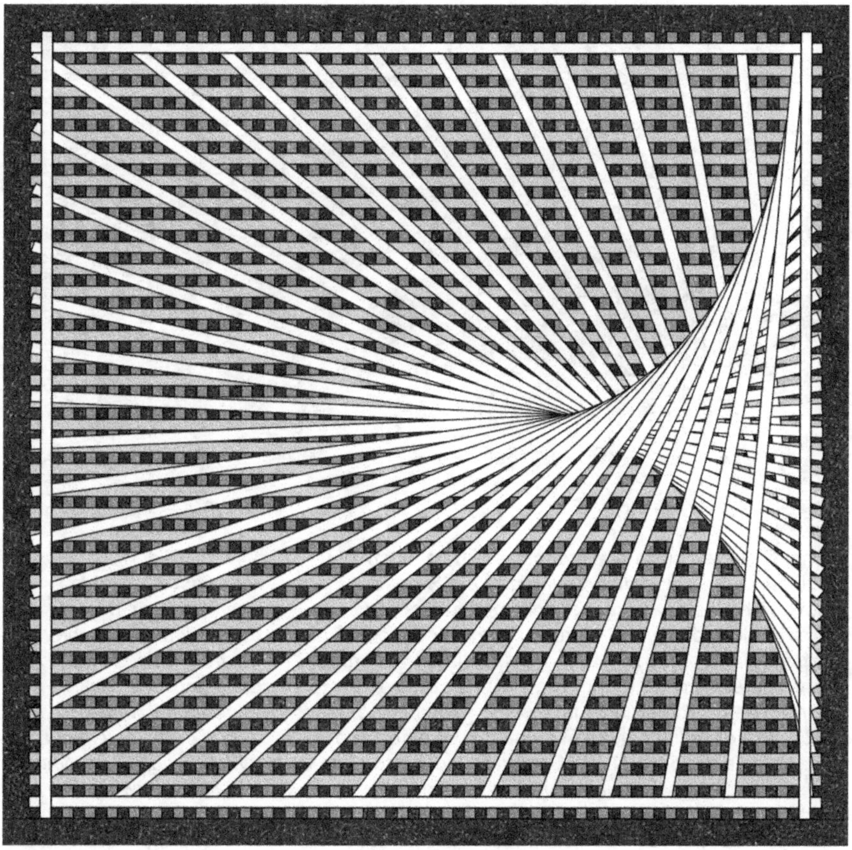

Organic Cube was built in Østre Anlæg, a garden in Copenhagen as part of an International Wood Festival in 2009. It is a 3.2 metre (10' 6") cube of timber laths within which more laths are arranged in a three-dimensional hyperbolic paraboloid swirl. The cube is sculptural – in that it is interesting to look at (above) – but it is also architectural – in that it identifies places for interaction. It can be entered as a 'cave', or climbed as a 'rock face'.

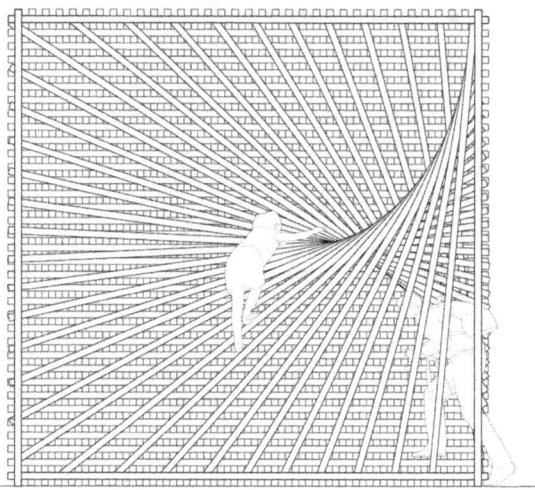

CURVES FROM STRAIGHT LINES
non-uniform rational B-spline (NURBS)

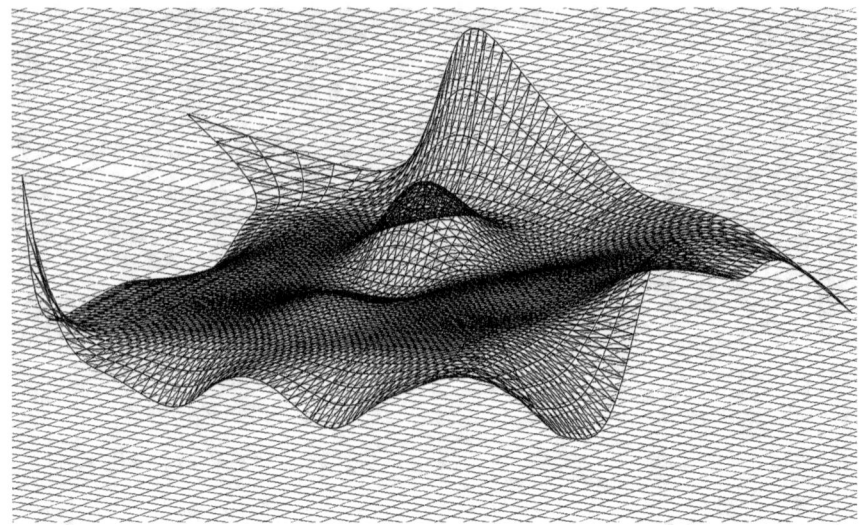

Complex curved surfaces and forms can be intuitively moulded within a computer using a NURBS surface (above). This uses a matrix of polygons and control points to allow designers to mould complex shapes. The polygons and control points move and distort according to parameters and algorithms set in the software coding. The NURBS matrix is a web of straight lines, the density of which can be stipulated by the designer. After manipulation the NURBS matrix is converted into a curved surface according to what is called the spline function, which generates smooth curves similar to those produced when a flexible material in the real world is bent, twisted and stretched in different directions (see page 6). Manufacturers and contractors can use the same computer models to calculate and make the elements needed to construct such complex curves as real buildings.

With the advent of sophisticated computer software for three-dimensional design and manufacture the constraints that impeded, for example, Erich Mendelsohn's attempt to create a curvaceous architecture of the future (see his Einstein Tower on page 45) were greatly lessened. Since the builders of Stonehenge used pegs and rope to layout their circles and horseshoe ellipses, architectural curves have been 'parametric'. But the term has become mostly applied to the algorithms used in software for the three-dimensional manipulation of curves. Such software has also enabled architects to design and contractors to build much more complex forms than could ever be achieved by adherence to the restraints of the geometry of making.

A NURBS surface allows a designer to create complex curved forms by manipulating a matrix of straight lines three-dimensionally within a computer. I generated the above surface within a few minutes using Autodesk 3ds Max software. It started with a flat surface which I pulled, pushed, stretched... The image can also be rotated to be viewed from different directions. Surfaces produced can seem like natural forms in the real world: rippling fabric; rolling ocean waves; the smooth interiors of sea shells...

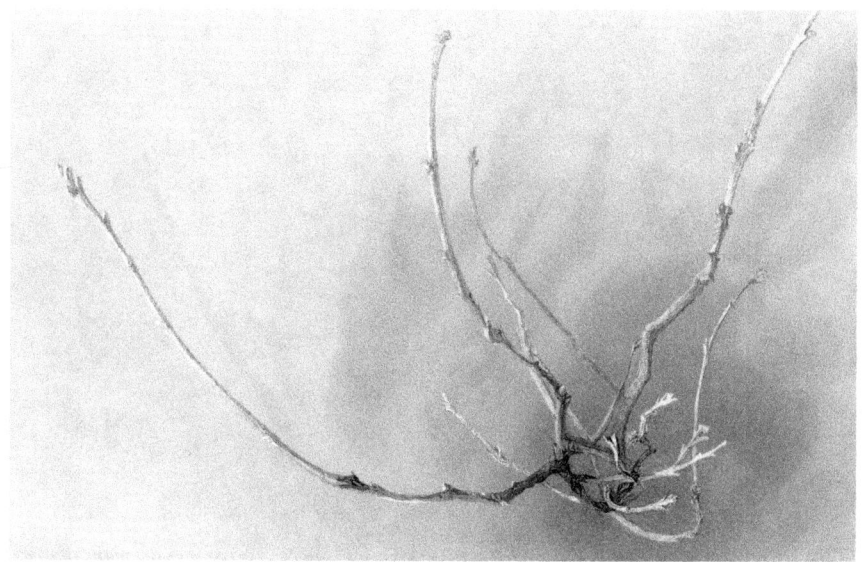

CURVES FROM NATURE

In this drawing, which he titled 'The Dryad's Waywardness', John Ruskin studied the ways in which branches grow in curves to avoid one another. Ruskin believed, almost as a form of worship, that human creativity should follow principles laid down by Nature. It is understandable that, since ancient times, architects have looked to natural creations – plants, animals, geology... – to find inspiration for the introduction of curves into their work. They have also sought out the underlying principles by which the beauty of natural curves is generated and according to which it might be emulated.

THE CORINTHIAN CAPITAL
inspired by a natural occurrence

'The third order, called Corinthian, is an imitation of the slenderness of a maiden; for the outlines and limbs of maidens, being more slender on account of their tender years, admit of prettier effects in the way of adornment. It is related that the original discovery of this form of capital was as follows. A freeborn maiden of Corinth, just of marriageable age, was attacked by an illness and passed away. After her burial, her nurse, collecting a few little things which used to give the girl pleasure while she was alive, put them in a basket, carried it to the tomb, and laid it on top thereof, covering it with a roof-tile so that the things might last longer in the open air. This basket happened to be placed just above the root of an acanthus. The acanthus root, pressed down meanwhile though it was by the weight, when springtime came round put forth leaves and stalks in the middle, and the stalks, growing up along the sides of the basket, and pressed out by the corners of the tile through the compulsion of its weight, were forced to bend into volutes at the outer edges. Just then Callimachus, whom the Athenians called κατατηξίτεχνος for the refinement and delicacy of his artistic work, passed by this tomb and observed the basket with the tender young leaves growing around it. Delighted with the novel style and form, he built some columns after that pattern.'

Above is a drawing of my attempt to replicate the origins of the Corinthian capital. The basket is replaced by a flower pot, the roofing tile by a concrete block. The acanthus has grown to greater maturity than suggested in Vitruvius's version of the story (left). But the idea adopted by Callimachus is clear.

The intention is to use the natural curves of plants, petrified in carved marble, as an ornament and foil to the regular constructional geometry of a temple.

Vitruvius, trans. Hicky Morgan (1914) – *The Ten Books on Architecture* (1st C BCE), Book 4, 1960.

Corinthian capital from the tholos at Epidaurus

And here is a drawing of an early example of the Greek Corinthian capital. It is from a fourth-century BCE tholos (The Thymele) at Epidaurus. According to the second-century BCE Greek traveller Pausanias, the building was designed by an architect and sculptor called Polykleitos.

The earliest example of a Corinthian capital was in the Temple of Apollo at Bassae (page 29). It dates from 427 BCE. It seems to have had particular significance, being used for just one column positioned at the focal point of the interior (right). The other columns are Ionic (see page 112).

reconstructed interior of the Temple of Apollo at Bassae

CURVE

OTHER PLANT-INSPIRED CAPITALS
ancient Egyptian examples

Columns are made round in section so that they appear the same width from all directions (left).

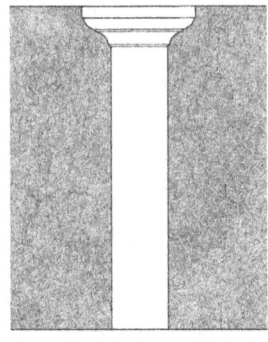

Since ancient times, architects have felt columns needed a capital to mediate between the shaft and the load supported (right). These often introduce curves.

In ancient Egypt, sculptors looked to plants – such as the palm, papyrus and the lotus – for inspiration for the design of capitals (below).

The capitals below are from the Temple of Isis at Philae, built in the fourth century BCE.

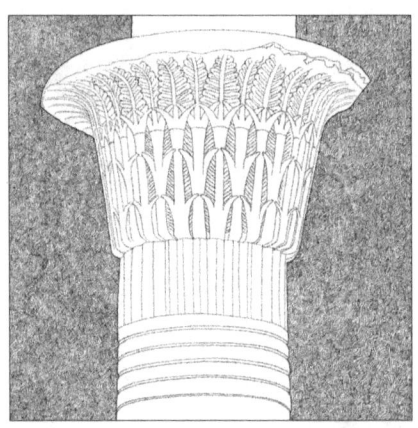
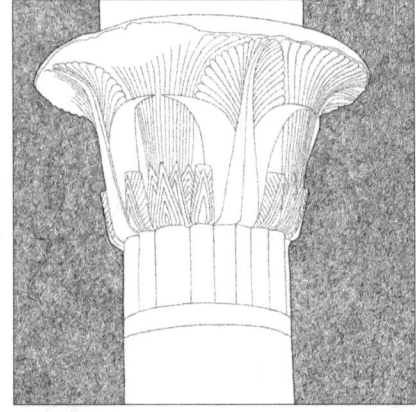
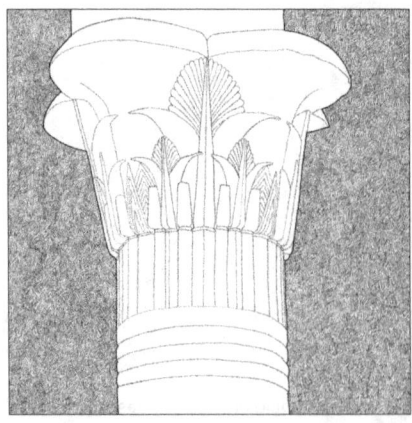
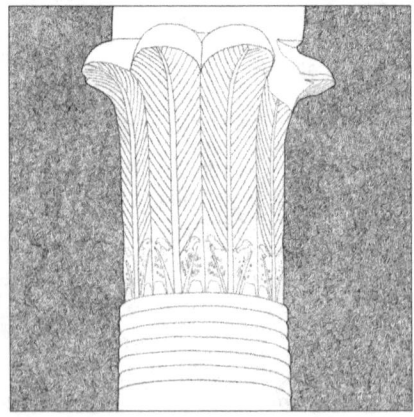

OTHER PLANT-INSPIRED CAPITALS
Gothic capitals from the Middle Ages

Ungewitter – *Lehrbuch Der Gotischen Konstruktionen* (1858), 1901.

SEASHELL SPIRAL
the source of the Ionic volute

The other columns inside the Temple of Apollo at Bassae (page 109) have Ionic capitals. The spiral volute of the ancient Greek Ionic capital is said to derive from spiral seashells (above). The method is to wrap string around the spiral of the seashell. As it unwinds (right), its end traces out the volute curve – a spiral that widens more quickly as it gets larger.

My attempt is illustrated in the drawing bottom right. Its curve matches, almost exactly, that of the Ionic column capitals of the fifth-century BCE temple of Parthenos at Neapolis (now Kavala in northern Greece), one of which is shown below.

Drawing a spiral using the sea shell illustrated above…
… with the resulting spiral below.

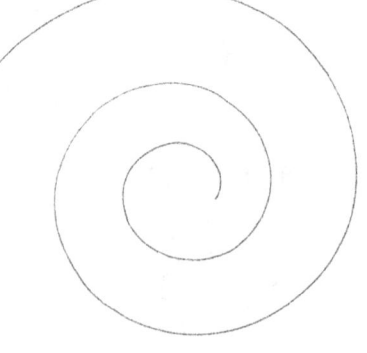

FECUND CURVES
curves of fruitfulness and regeneration

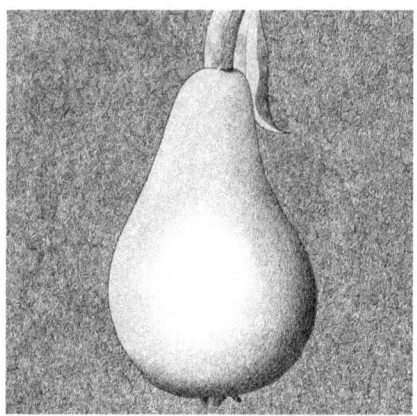 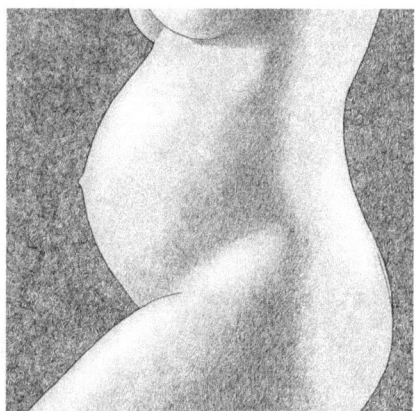

Curves can suggest fecundity. Fruit, eggs, pregnancy bumps... All are curved and suggest fruitfulness, renewal, procreation...

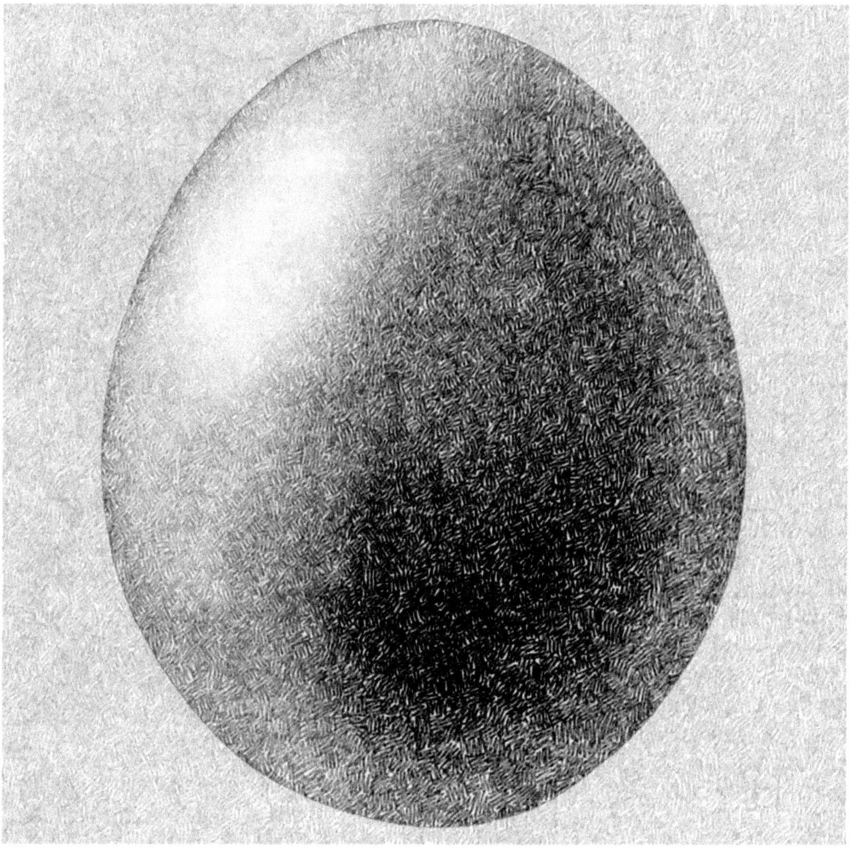

OVOID CURVES
eggs as symbols of origin, nativity, creativity

The egg has inspired the forms of many types of building. Its suggestion of birth and renewal makes it an apt symbol for religious buildings of all sorts.

But its inherent structural strength, which can resist forces of internal pressure as well as of gravity, also makes it suitable for industrial purposes.

The egg is one of the simplest, subtlest and most beautiful forms in nature. It is also structurally strong.

Symbolically, the egg represents nurturing and protection. From eggs comes new life.

It is understandable that architects around the world have sought to emulate the egg form in building. Many religious buildings, of different faiths, draw on this symbolism. It can also be applied to buildings that accommodate creative arts.

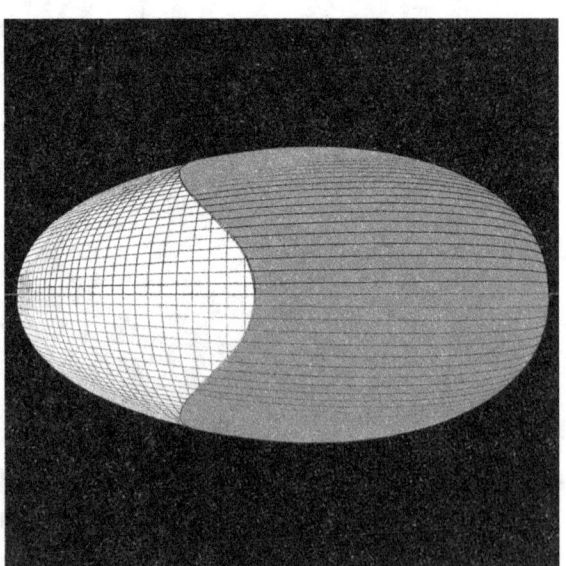

Reflected in its lake, the National Centre for the Performing Arts in Beijing, designed by Paul Andreu and built by 2007, appears as an egg... an egg that, presumably, nurtures creativity.

EROTIC AND SENSUOUS CURVES
the curves of human form

'She looked at a man because she liked the way the hair was tucked behind the ears, or she liked the question-mark line of a long torso curving at the shoulder and straight at the hip.'

Maxine Hong Kingston - 'No Name Woman', in *The Woman Warrior*, 1977.

We human beings are innately curvy creatures. Some of us are more attractively curvy than others. Our inborn appreciation of and instinctive reaction to the curves of the human form must condition our aesthetic appreciation generally.

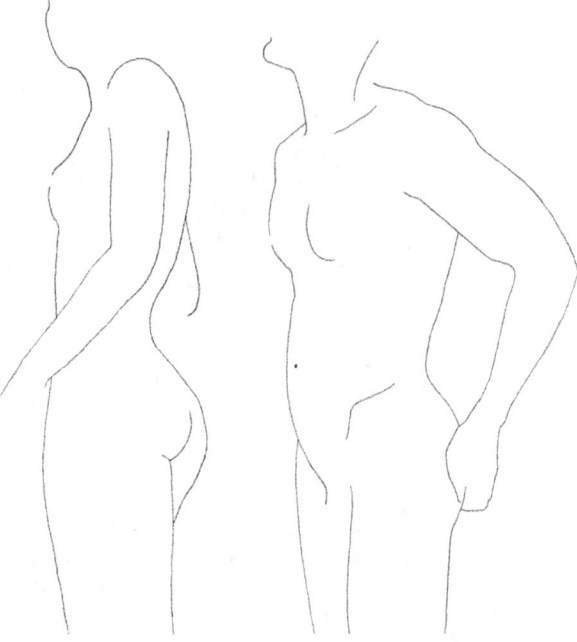

Sometimes human curves are transposed directly into architecture through sculpture (as in the caryatid above), sometimes more subtly.

We are predisposed to appreciate and respond to the curves of the human form.

Life drawing challenges the artist to get the curves right, not only in the lines of the body's profile but also in the shading of the flesh.

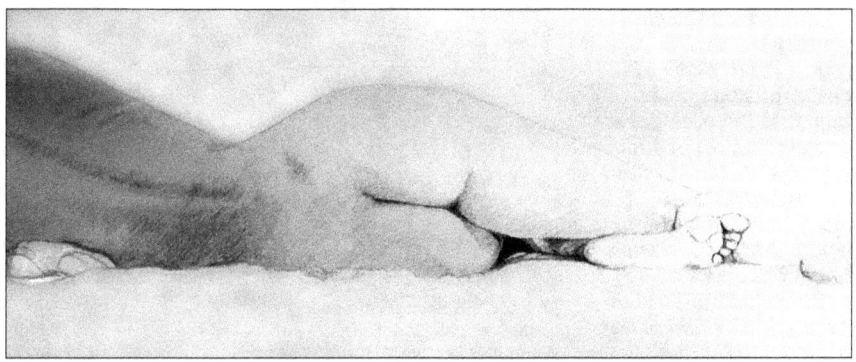

EROTIC CURVES
of landscapes and clefts

ARCHITECTURE IS THE ULTIMATE EROTIC ACT
CARRY IT TO EXCESS

Monica Bonvicini, 'Stonewall 3', 2002.

We are attracted to the refuge offered by the anthropomorphic recesses of caves and clefts in trees. Sometimes architects seek to emulate these with curvaceous architecture; as in Amâncio d'Alpoim Miranda 'Pancho' Guedes' design for a house which he described as a 'habitable woman' (below).

'Ladies and gentlemen, I present my first habitable woman; a vigorous, personal, erotic, obscure, suprafunctional, suburban citadel... An anthropomorphic wonder-house... A round-eyed house of cavernous passages built into the rock gaps, or swinging walls finding their own levels, of fluid ceilings and floors, of many stairs and steps, with lights staring and pouncing out of ceilings and walls; a house with a baby-house inside her. A pregnant building.'

Amancio d'Alpoim Guedes, quoted in John Donat, editor – *World Architecture* 2, 1965.

'Gournia's double hills... are so close and rounded that a more proper analogy would seem to be directly to the female body itself, and they do closely resemble the uplifted breasts of the "goddess of the horizon", topping her horns or crotch beneath... Indeed, at Gournia one has the inescapable impression that human beings are conceived of as children who lie upon the mother's body, enclosed by her arms and in the deep shadow of her breasts.'

Vincent Scully – *The Earth, the Temple and the Gods* (1962), 1979.

The landscape is commonly considered as female; its curves, mounds, clefts... interpreted as parts of the body of mother earth. That interpretation translates into architecture too, into the interventions we make to our surroundings in order to amend them to our needs and desires and to try to meet our aspirations. Architecture, at its roots can be anthropomorphic; and, being so, can clearly include allusions to the sexual and erotic characteristics of the human form. These prompt some of our most powerful emotional reactions and responses; it would be surprising if they did not inform our creative responses to the world in which we live, as in prehistoric burial mounds and contemporary earth sheltered houses (below).

Bryn Celli Ddu (left) is a burial mound on the Isle of Anglesey off north Wales. It is around five thousand years old; maybe a few hundred years older than the ancient Minoan settlement at Gournia described by Scully in the quotation at the top of this page. Like many such burial mounds, its womb-like forms suggests that its late-Neolithic builders also interpreted their landscape as mother and amended it with suitably feminine curves.

Bryn Celli Ddu, Neolithic burial mound

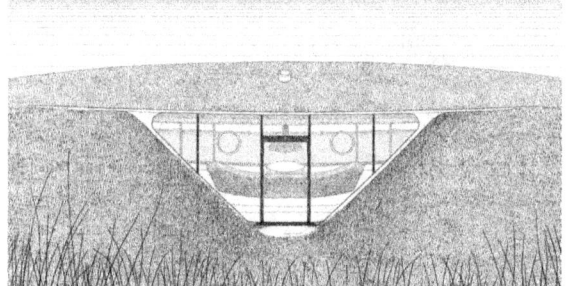

Malator, Future Systems, 1998

Something similar happens in the case of the small holiday house Malator (left), built on the coast of Pembrokeshire in the 1990s and designed by Future Systems. It too has an entrance like an entrance into a womb, though, unlike a burial chamber, this 'womb' has a view to the setting sun across the Irish Sea.

See also Future Systems' 1998 entrance to the Comme des Garçons store in New York on page 91.

EROTIC CURVES
of openings and erections

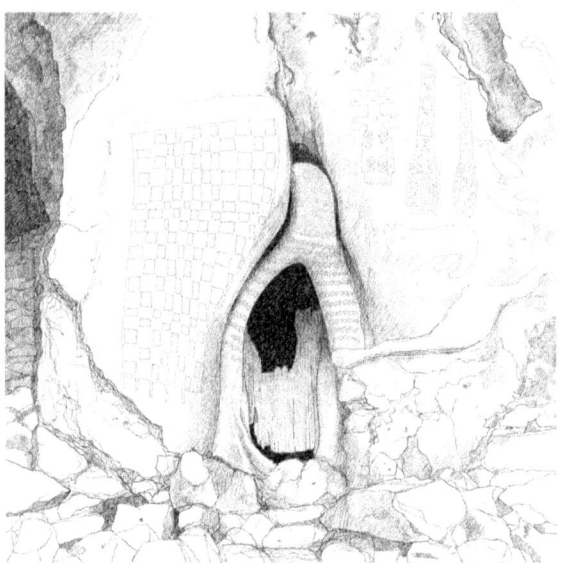

The allusion of a cave opening to the entrance to a womb – the womb of mother earth – is made explicit in this carved entrance to a Dogon shrine to the sacred feminine in Mali (above).

Of course, whether intended by their architects or not, there are instances of curved architecture that have, in the popular imagination, been classified as 'phallic' (below). On the left is another project (in this case unrealised) by Future Systems – Green Bird, 1998 – which would have been taller than the Empire State Building; while on the right is 30 St Mary Axe, London, designed by Foster + Partners and completed in 2004.

In 'Imponderabilia' (1977, above) Marina Abramovic and Ulay transformed the hard rectangular opening of a doorway with the soft warm vulnerable curves of their own naked bodies. It was an architectural intervention. Their impassive occupation could not do other than instil the opening with an erotic charge for those squeezing through.

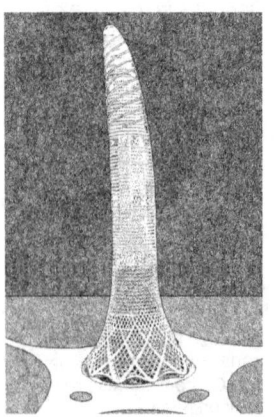 

Curved buildings might be interpreted as having 'male' characteristics as well as female.

HAPPY AND SAD CURVES
smiles and grimaces

In common with most children my granddaughter enjoys trick drawings. One she likes is a face which, manipulated in different ways, looks sad or happy. The difference is a matter of curves. It illustrates the powerful effect curves can have on our perception and empathy with emotional expression.

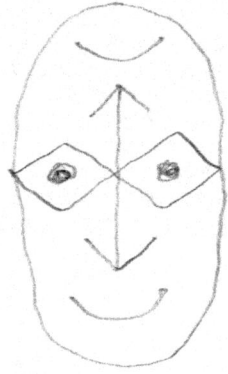

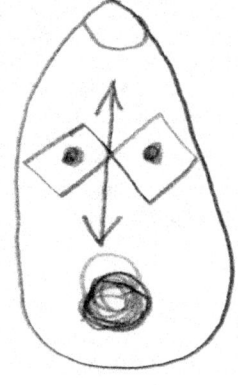

Turn the book upside down and observe the dramatic change in expression. Look particularly at the eyes, which cannot change graphically but appear to do so emotionally.

Place your finger or thumb over one side of the mouth and then the other. Observe a similarly dramatic change to that when the previous image is turned upside down. (Watch the eyes.)

This is a drawing my granddaughter invented for herself. Again the trick is to turn the book upside down and observe the change in facial expression, this time from horror to abject misery.

The French pointillist painter Georges Seurat famously believed that upward curves evoked joy. In 'Can-Can' ('Le Chahut', 1890) the mood seems more sinister than joyous, rather like a fixed smile or rictus.

The upward curve of the roof of Le Corbusier's Ronchamp chapel adds to the building's expression of joyous aspiration.

ENTASIS
a muscular curve

Entasis is the swelling of columns noticeable in many ancient temples (from Egypt to Rome and in other parts of the world) and replicated in revived classical architecture from the Renaissance through to the present. It is generally thought that the architects of ancient Greece developed entasis for optical/aesthetic reasons – to make the profile of a column *look* straight, straighter than a straight profile would. But it seems plausible that they also did it for expressive reasons, to mimic the swelling curves of muscles and make the column appear strong.

Through the Renaissance, as probably also in Classical times, there was discussion amongst architects as to whether there is a rule for the construction of the entasis curve, and what that rule might be.

The columns of the first Temple of Hera (the so-called Basilica; c.550 BCE) at Paestum have particularly pronounced entasis (below). The columns of the adjacent second temple (which may be seen in the background; c.500 BCE) have a more subtle curve.

The swelling of a column – entasis – mimics the swelling of muscles, and hence expresses strength.

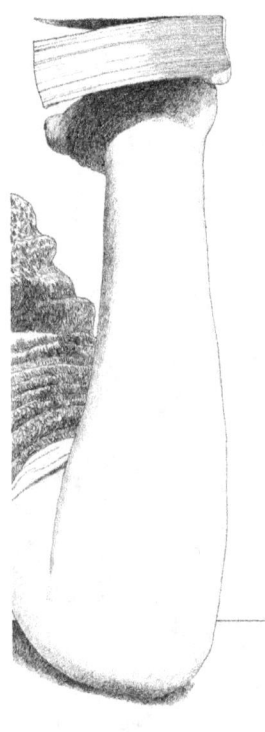

Through history architects have put forward different ideas for how entasis should be established. In this diagram (below left), drawn in the early twentieth century by Rhys Carpenter, the entasis apparent on the columns of the Temple of Mars Ultor in Rome is compared with the method proposed in the sixteenth century by the Italian architect Giacomo Barozzi da Vignola. (It is in Hickey Morgan's 1914 translation of Vitruvius.)

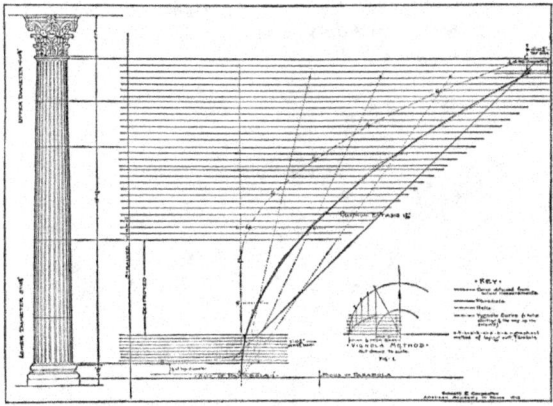

Carpenter's analysis suggested entasis was, in the instance studied, parabolic (and subtly different from Vignola's rule). In the late 1970s, German archaeologists discovered construction drawings scratched onto a wall of an ancient Greek building at Didyma (in modern Turkey). One of these, replicated below, gives a method for establishing entasis. It shows half a column, diameter 2m, 'squashed' to 1/16 its proper height. The curve to the left is an arc of a circle with its centre level with the point at which the column face first becomes vertical, and with a radius of 3m. When the diagram is stretched back to its proper full height the arc becomes elliptical and fits the profile of columns at Didyma (right).

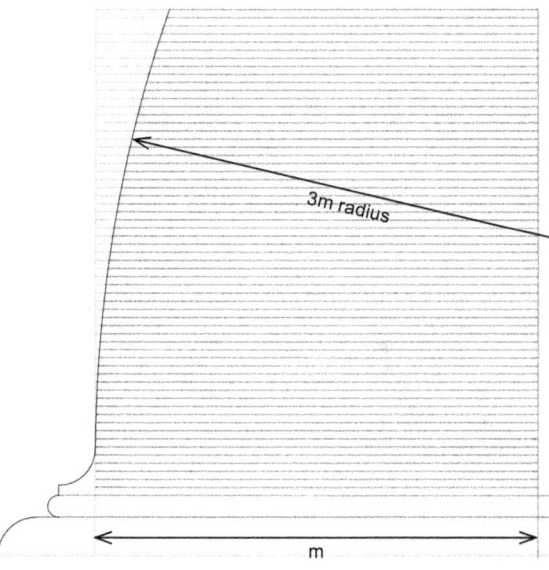

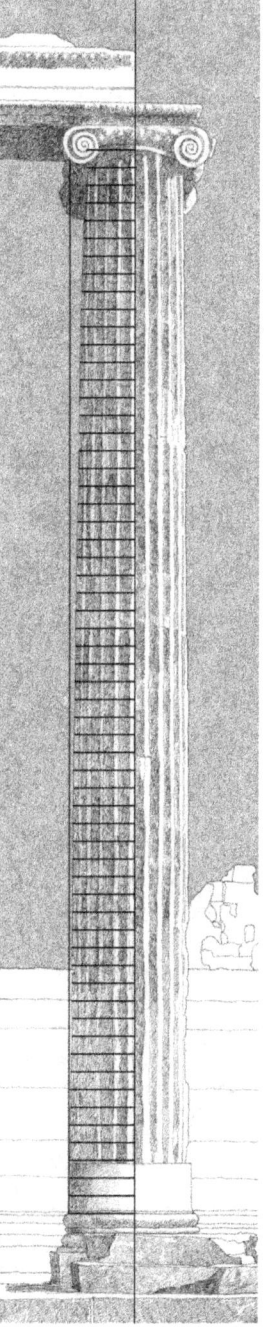

TRACERY OF TREE BRANCHES
filtering the light from heaven

'For bright is that which is brightly coupled with the bright,
And bright is the noble edifice which is pervaded by the new light...'

Abbot Suger, trans. Panofsky
– *On the Abbey Church of St.-Denis and its Art Treasures* (12th C), 1946.

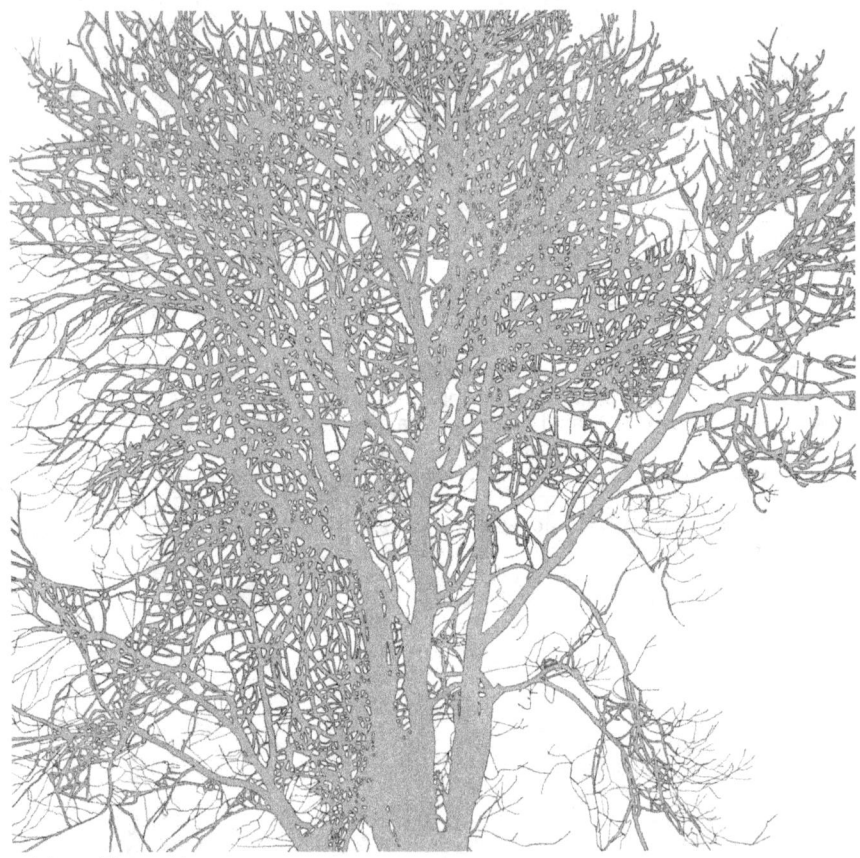

ash tree of the world

The grandeur of great trees has numinous power even in the winter when denuded of leaves. They are maternal and protective, and have the presence of animistic gods. Part of their beauty derives from their silhouette against the bright sky. The intricate curves of their twigs and branches create a sheltering canopy, the tracery of which filters the light from the heavens above. It is understandable that Gothic architects drew inspiration from their form. The curves of the tracery of branches are formalised in the curves of the tracery of windows in Gothic churches.

The tracery of a great tree's bare branches is silhouetted against a bright winter sky.

St Mary, Snettisham, Norfolk, fourteenth century

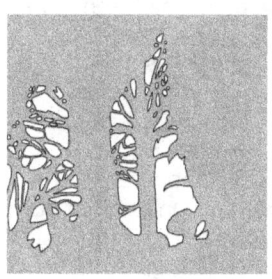

The east window in the Church of St Mary, Snettisham, (above) is filled with stained glass (not shown). Its tracery is more sinuous in its curves than many Gothic church windows. But these curves only make the window more reminiscent of their inspiration. Like all Gothic tracery, it is a formalised version of the branches of great trees that filter the light from heaven. This lends it metaphorical power – 'the tree of life' – as well as giving it an aesthetic power that derives from the tracery's curving intricacy and the way in which sunlight illuminates the church interior in the morning.

TENDRILS
Castel Béranger, Paris, Hector Guimard, 1898

Castel Béranger, Paris

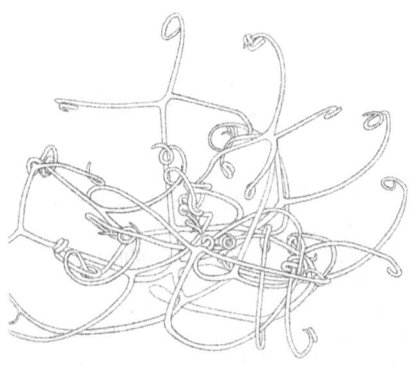

Plant tendrils grow in a tangle of exploratory curves.

The gateway illustrated above is the entrance to the Castel Béranger in Paris seen from inside. It was designed by French architect Hector Guimard and built in the late 1890s. At the time, architects and designers across Europe were interested in the decorative use of curves taken from nature, especially growth curves. Guimard's design for this gate seems influenced by the sorts of tendrils found on climbing plants such as vines, peas, clematis, wisteria… Guimard used such apparently free curving 'natural' lines in many of his works including the early Paris Metro stations.

LIGHT REFLECTING OFF WATER
Casa Milà, Barcelona, Antoni Gaudí, 1912

Casa Milà, Barcelona

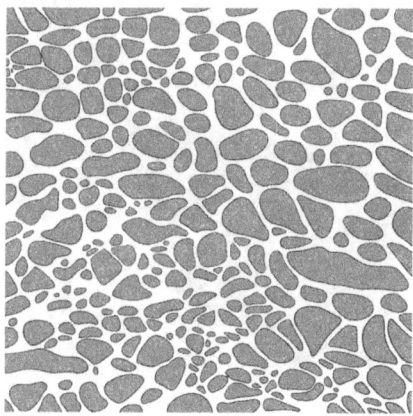

Light reflects in dancing curves on the surface of water.

In the same period, the Spanish architect Antoni Gaudí (see also pages 78–9) was experimenting with free curves. Above is a drawing of his street level gates into the Casa Milà central courtyard (1912), again seen from inside. These gates appear to evoke the looping reflections of sunlight off the surface of rippling water (left). Gaudí too, like Guimard, used the curves of plants and other natural phenomena – erosion, waves breaking, shells, seed pods, ant hills, stars, wings, strange creatures, armadillo carapaces, tortoise shells, fish skeletons (see page 128)... – in many of his works.

PEBBLES
Aquarium (project), Batumi, Henning Larsen, 2010

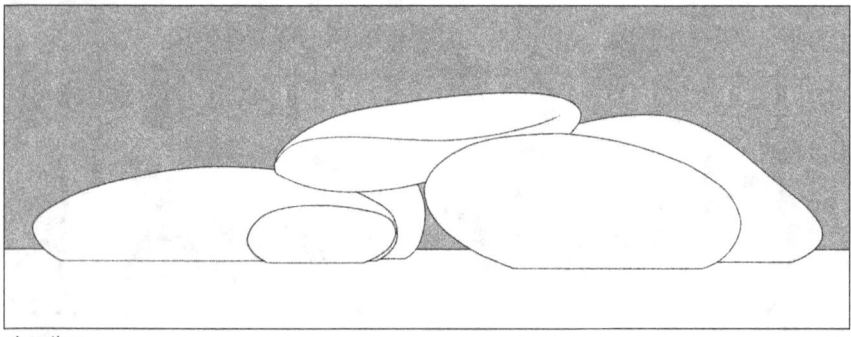
elevation

In 2010, Danish architectural practice Henning Larsen Architects won a competition to design a new aquarium for the Georgian city of Batumi on the Black Sea. Larsens chose to base their design on a carefully composed simple pile of pebbles from Batumi's beach.

It was not clear from the competition drawings what the structure of the pebbles would be. The design has not yet been built.

plan

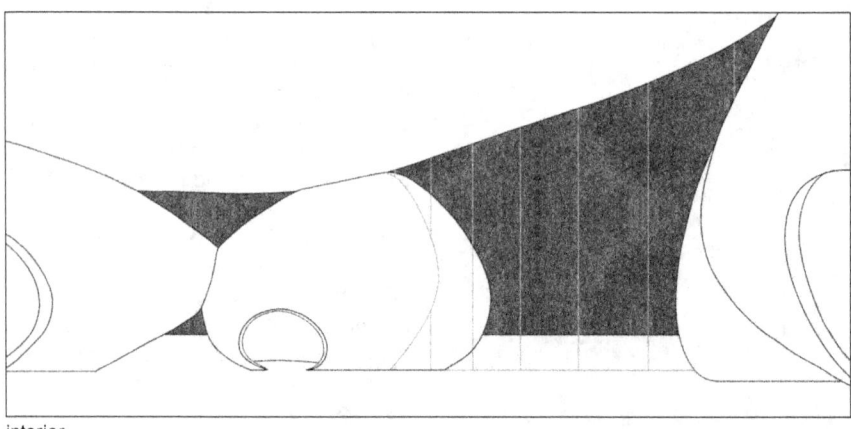
interior

BONES
Henry Moore; Doha Villa, Kathryn Findlay, 2004

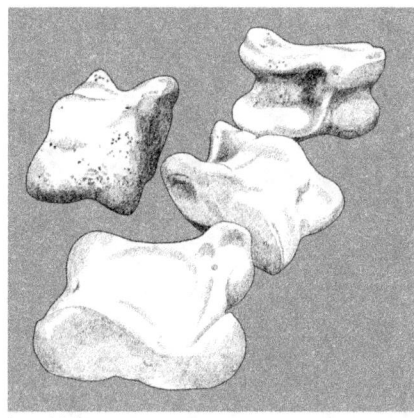

knucklebones

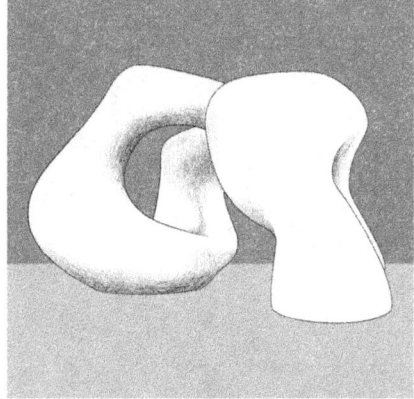

Henry Moore, 'Two Large Forms', 1960s

The British sculptor Henry Moore used bones as inspiration for his sculptures (above). He did not want bone sculptures, but wanted to emulate their curved forms.

Kathryn Findlay was not directly inspired by bones when she designed the Doha Villa (below as a model; the house was started but not completed and has now been demolished). But the form of the house benefited from the same kind of abstract curves, explored through the use of computer software.

Kathryn Findlay also designed the Truss Wall House (see page 60). That earlier, and similarly curvy, house was designed largely without the help of a computer.

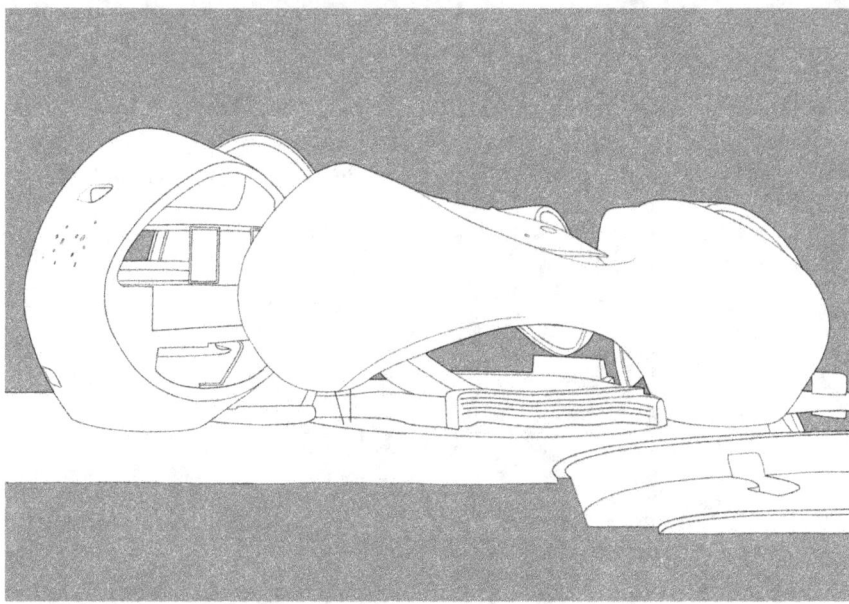

Kathryn Findlay, Doha Villa, 2004

CURVE

FISH SKULL
Casa Batlló, Barcelona, Antoni Gaudí, 1904

a balcony from the Casa Batlló

Antoni Gaudí avoided straight lines in his architecture. He used curved lines inspired by nature.

The balconies of the Casa Batlló (above), for example, are inspired by the curves of fish skulls (below).

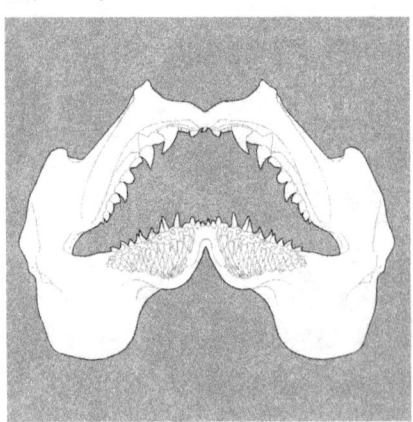

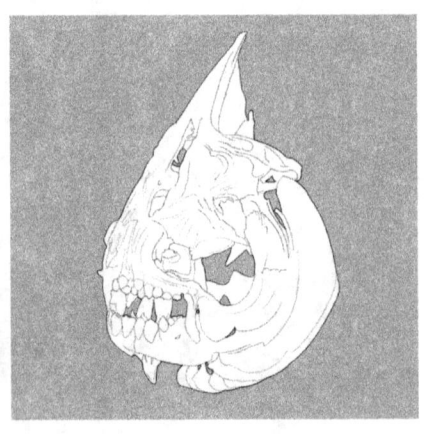

HUMAN SKELETON
Farkasrét Mortuary Chapel, Imre Makovecz, 1975

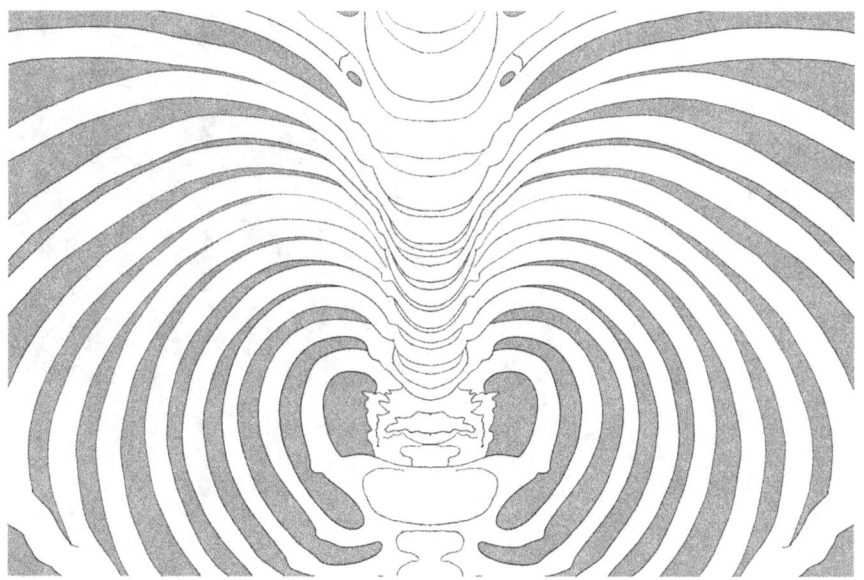

Imre Makovecz was a Hungarian architect who created fantastic anthropomorphic and zoomorphic forms in his architecture, mostly in timber. Being inside the mortuary chapel at Farkasrét Cemetery in Budapest is like being inside a human chest. The structure is composed of symmetrically curved timbers meeting in a spine that runs down the ridge, just as in a ribcage.

Notice that the door of the chapel is where the pelvis would be in a human skeleton – the portal through which we are born into the world and, here, leave.

CAVES
emulating the curves of erosion and excavation

Scoured out by the natural flow of water, the shapes of caves are irregular curves. Architects' attitudes to such curves vary from trying to negate them, through exploiting them, to emulating them in constructed form.

Peak Cavern, Derbyshire, in an 1830s woodcut

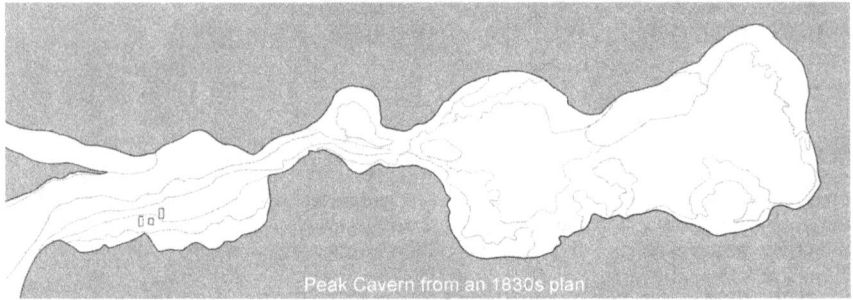

Peak Cavern from an 1830s plan

Peak Cavern (above) is a natural cave. The drama of its size and eroded surfaces contrasts with the smallness and squareness of the cottages of the rope-makers who once lived and worked there.

In some rock-cut architecture, such as the Elephanta cave temples in India (below), the aim has been to introduce the orthogonality of conventional architecture into the carved out spaces within the rock. In this way the naturally curved surfaces of a cavern are largely negated and replaced by straight flat walls, straight lines and axes (as in architecture built orthogonally).

The cave temples of Elephanta (right) are carved out of the natural rock. But the architecture of the spaces largely follows the orthogonal geometry of conventionally constructed, rather than excavated, architecture.

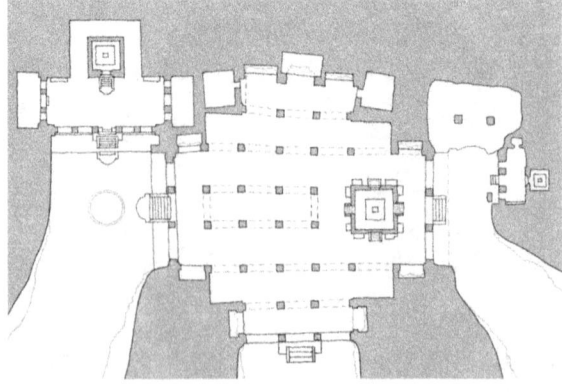

Whereas some other, even more ancient, rock cut temples exploit and develop the freedom of excavating space by creating labyrinthine caverns with curvy walls and pathways between.

The Hypogeum in Malta dates from probably over three thousand years ago. Its subterranean spaces have the amorphous form of a natural cave system, but developed and extended for ritual purposes.

plan of the Hypogeum

The ancient culture of Malta also built great temples above the surface of the earth. But in doing so they sought to emulate the indistinct form of caves. The temple on the right is Hagar Qim. Although it has some axial pathways, its main spaces have curved walls. Without light these curved spaces would have the indistinctness of natural caves.

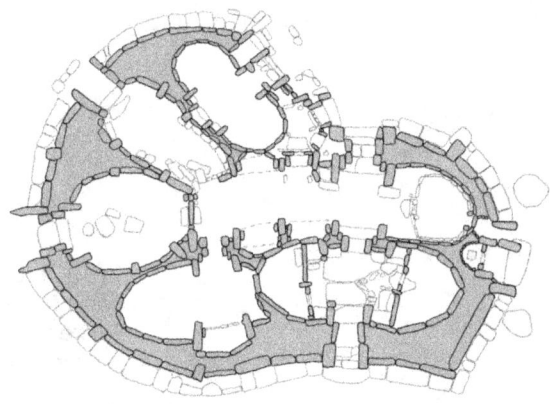

plan of Hagar Qim

Architect Richard England was influenced by these forms when designing Manikata Church on Malta in the 1960s (right). The curving walls and absence of corners contribute to its cave-like interior.

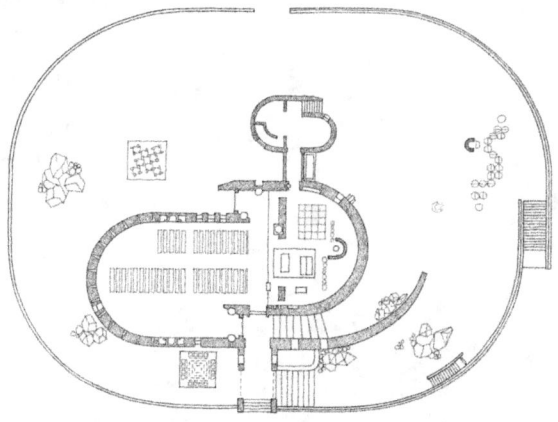

plan of Manikata Church

CAVES: INTANGIBLE WALLS
Ronchamp Chapel, Le Corbusier, 1954

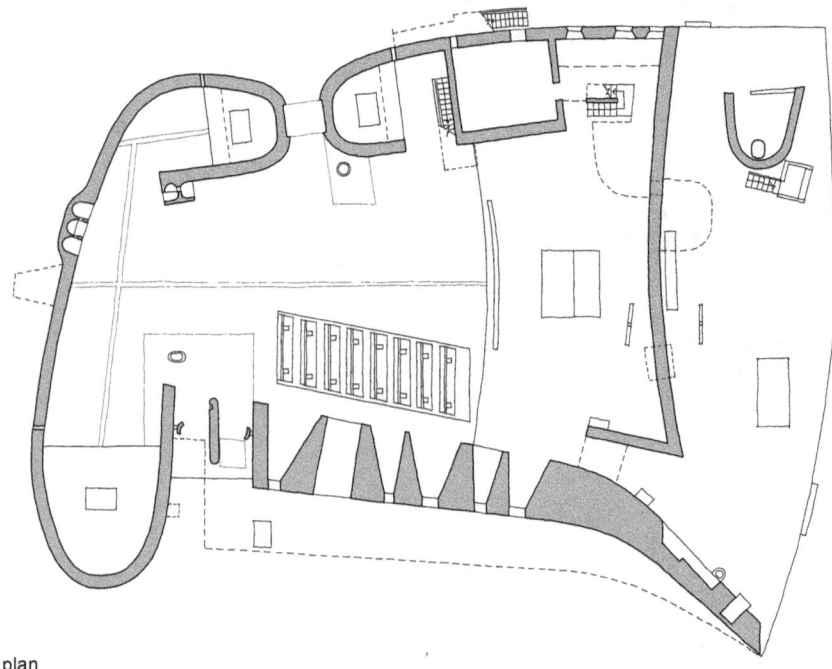

plan

a top-lit side chapel

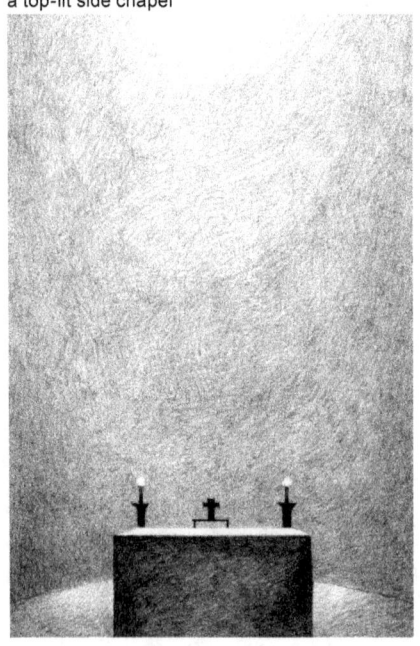

In 1954, Le Corbusier used curved walls for his chapel at Ronchamp to create a cave-like interior without distinct edges or boundaries.

Curved surfaces, without distinct corners or other features that might provide reference points, can create a sense of intangibility. They are solid, but they can seem like air. In the dark recesses of caves the smoothed curved surfaces of the rock can evoke infinite blackness. Lit obliquely, as in the small side chapels of the Ronchamp chapel (left), wall surfaces seem like atmosphere. Such effects add to the power of curved walls and their contribution to the emotional experience of buildings.

Evenly textured curved surfaces can dematerialise, appearing to be atmosphere rather than solid material.

BUILDING CAVES
Silkeborg Museum, Jørn Utzon, 1964

In the early 1960s, the Danish architect Jørn Utzon was asked by the painter Asger Jorn to design a gallery for his work. Though the gallery was not built in this form, Utzon's initial designs envisaged the gallery as a series of subterranean chambers with cave-like concave walls and lit from openings to the sky. The section and plan of this design are below. Notice too that the design includes a complex system of internal curved ramps.

According to Richard Weston's authoritative book on Utzon*, the design of the Silkeborg Museum was influenced by his knowledge of subterranean Chinese structures. The section and plan below illustrate the examples given in Weston's book.

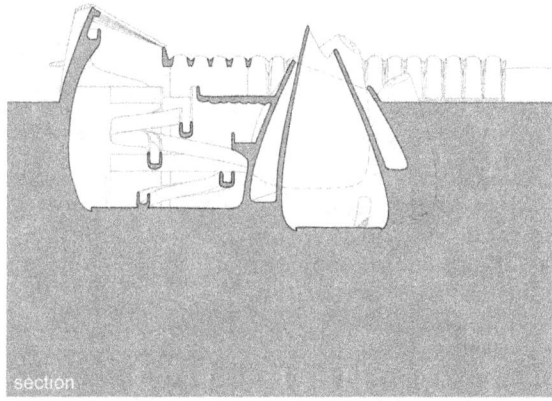

section

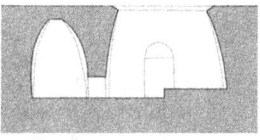

section

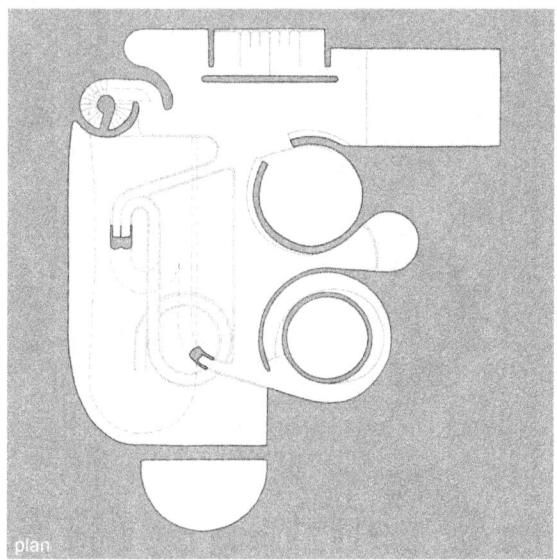

plan

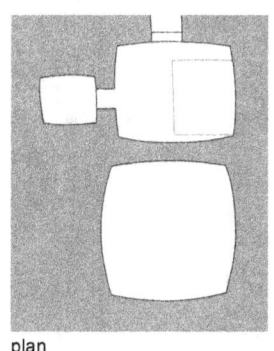

plan

* Richard Weston – *Utzon*, 2002.

CLOUDS
Bagsværd Church, Jørn Utzon, 1976

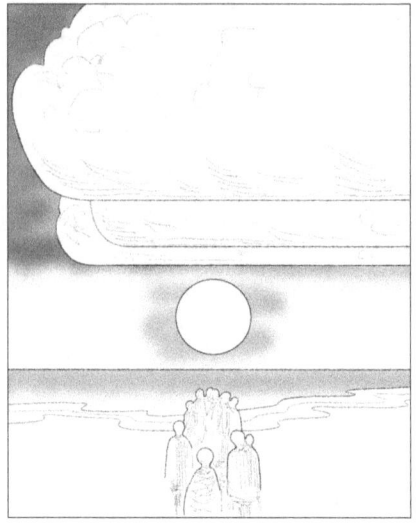 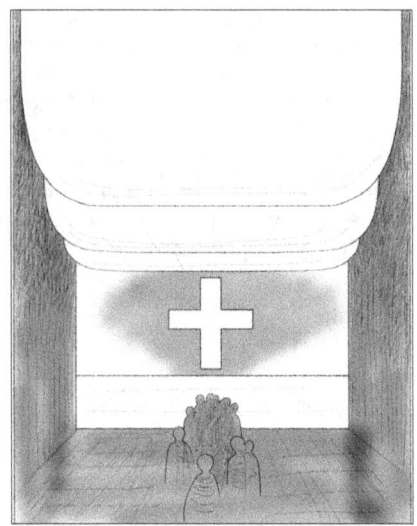

The ceiling of Utzon's Bagsværd Church in Denmark is composed of curved panels (see the section below). The idea was inspired by Utzon seeing regular banks of clouds building in the sky over a Hawaiian beach. The curved panels in the church emulate those clouds, evoking an inner landscape with a formalised sky moulded in billowing white concrete shells suspended from an outer structure.

To illustrate his idea Utzon did two sketches (similar but better than mine above). He showed the clouds over the Hawaiian sea and, alongside, the billowing ceiling he wanted to create in his church. (I have added the sun in the beach sketch.)

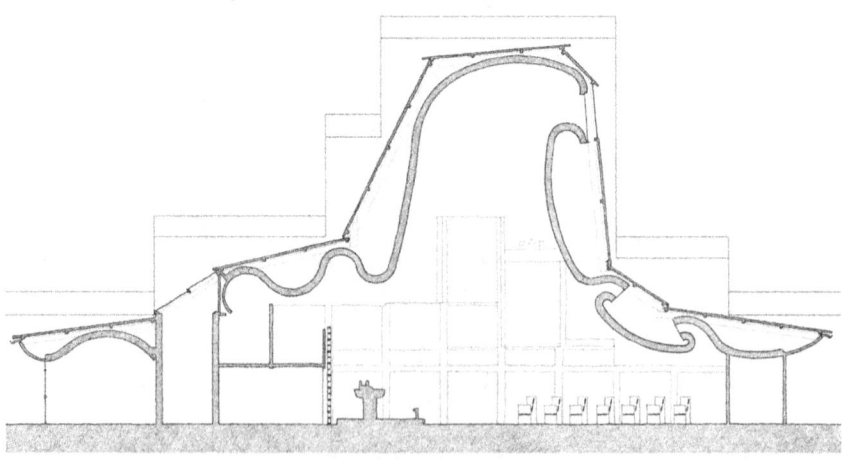

The section through Bagsværd Church shows the curved panels forming its 'sky'. These 'clouds' are enclosed inside a construction composed of straight elements.

See Richard Weston – *Utzon*, 2002.

EMULATING A NATIONAL ATMOSPHERE
the curves of Alvar Aalto

Aalto is a Finnish word for wave, so it is not surprising that Finland's most significant twentieth-century architect was interested in exploring the various roles of the curve in architecture. Most of the ways in which Alvar Aalto used curves were to emulate characteristics of the Finnish landscape in architectural form.

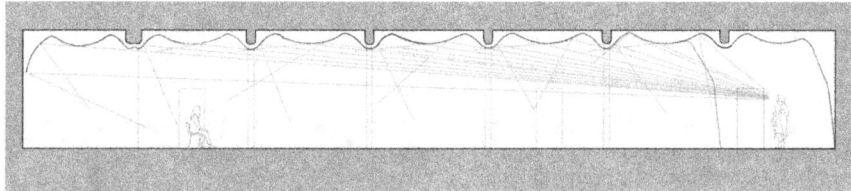

In the late 1920s and early 1930s Aalto designed the Viipuri (Russian name Vyborg) Library. Its accommodation included a long narrow lecture hall, the acoustics of which were likely to be problematic, partly because of beams across the ceiling. So Aalto designed a wavy timber false ceiling (above), wrapping under the beams, which helped the acoustics but also gave the space something of the character of a natural cave.

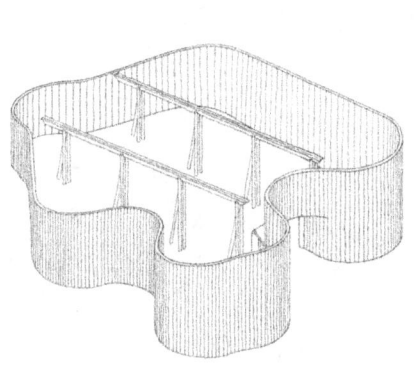 

In 1938 Aalto designed a Forestry Pavilion for an agricultural exhibition in Lapau. The plan has the characteristic wavy form that Aalto also used for his famous glass vases (right) but with vertical walls clad internally and externally in timber. This enclosed curved loop derives from forms in the Finnish landscape. The pavilion has the appearance of tight-packed trees around a forest clearing or pond.

The pavilion's flat roof was supported on straight timber trestles (above).

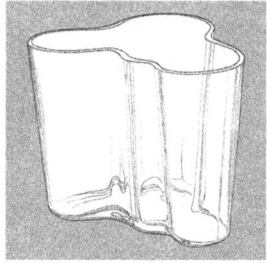

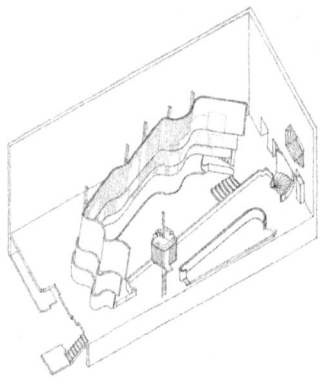

Aalto designed the Finnish Pavilion for the Universal Exhibition held in New York in 1939. Contained within a rectangular box (above), Aalto's design featured a wavy tiered wall, constructed in timber, used for the display of images of Finland. This dramatic wall was intended to evoke the Northern Lights, or Aurora Borealis (right).

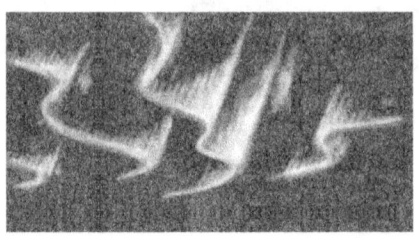

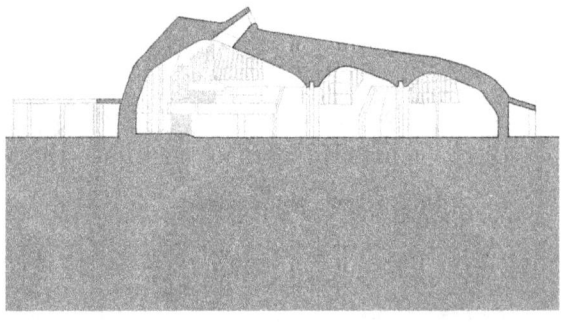

And Aalto's design for Vuoksenniska Church in Imatra was completed in 1958. The section and plan (right) illustrate that its interior space, with white walls, is like the scooped out form of an ice cave – a refuge from the wintry wastes. (This distinctively shaped nave can be divided by sliding walls to accommodate different sizes of congregation.)

See also Snøhetta's Norwegian Wild Reindeer Centre on page 38.

BIRDS IN FLIGHT
TWA Terminal, New York, Eero Saarinen, 1962

Architects have found many forms in nature to emulate. So far, in this section, we have seen examples inspired by plants, shells, trees, caves, clouds, lakes, forest clearings, Aurora Borealis and ice caves.

Appropriately for an American airline, Eero Saarinen, like Aalto a Finnish architect, found inspiration in the form of a swooping eagle for the design of his Trans World Airlines Flight Center at John F. Kennedy International Airport in New York (above).

The curved forms of the building are made possible by the use of reinforced concrete shell structures (as in the Brynmawr Rubber Factory on page 84). But here they are used not merely for practical reasons but to convey ideas – the combined ideas of flight and the United States of America.

The curved forms of the TWA Terminal emulate the dynamic energy of an eagle swooping to grasp its prey (below).

COBWEBS
German Pavilion, Montreal Expo, Frei Otto, 1967

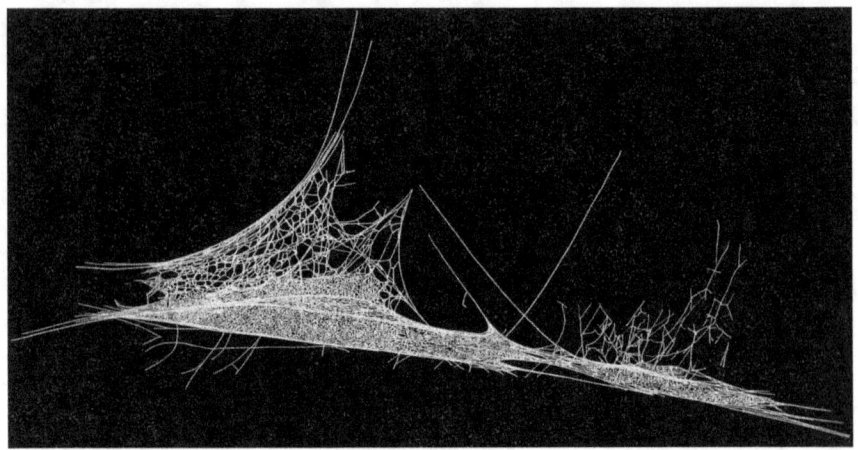

Delicate, beautiful, curvaceous forms in nature fascinate the creative imagination. They offer examples from which to learn. They provide inspiration that generates new ideas. They present challenges to aspiration. Maybe those challenges cannot ultimately be met but they raise the bar of human creativity.

In the 1960s, the German architect and engineer Frei Otto began experimenting with tensile structures reminiscent of cobwebs. In 1967 he was asked to provide the design for the German Pavilion at Expo'67 in Montreal. The drawing below shows the underlining web-like structure before the weatherproof membrane was fitted.

Cobwebs are delicate nets of fine gossamer stretched from anchorage points and curved under tension. Even scruffy broken ones possess their own delicate beauty.

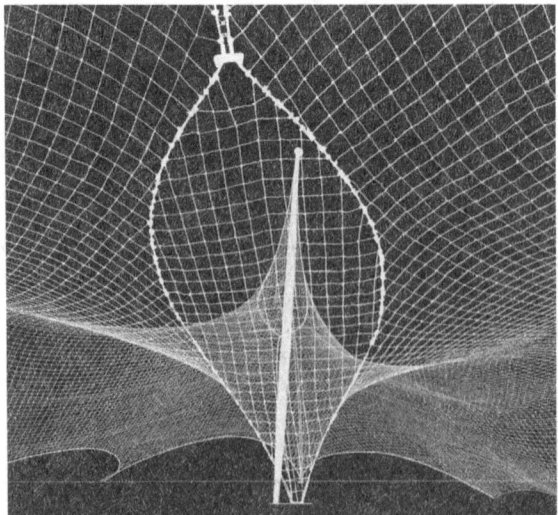

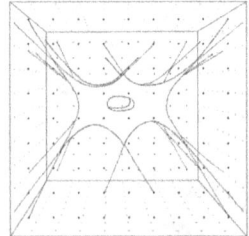

Otto was awarded the RIBA Gold Medal in 2006. Students of architecture continue to be fascinated with the forms achieved with tensile and membrane structures. Above is a study model produced by a student in the Welsh School of Architecture.

AERO(HYDRO)DYNAMIC CURVES
movement in relation to fluids – air and water

Some people's definition of architecture might include a provision that it does not move. This is not necessarily so, but nevertheless a general presumption that buildings are static objects and that therefore optimisation of acceleration and progression through fluids is irrelevant to architecture.

From the natural form of creatures that fly and swim, and from the designed form of vehicles, we know that curves are important to aero- and hydrodynamics. We understand such curves as pragmatic – as enhancing efficiency of movement – but we also appreciate them as beautiful.
Whether for pragmatic or aesthetic reasons, the architects of static buildings have been influenced by the subtle curves associated with aerodynamic and hydrodynamic profiles.

Aeroplanes, birds, cars, sharks, submarines... all have beautiful forms that we appreciate as deriving from aerodynamic or hydrodynamic efficiency. Architects might seek to emulate that efficiency, but they are also seduced by the aesthetic and metaphorical possibilities.

AERO(HYDRO)DYNAMIC CURVES
towers and Oscar Niemeyer

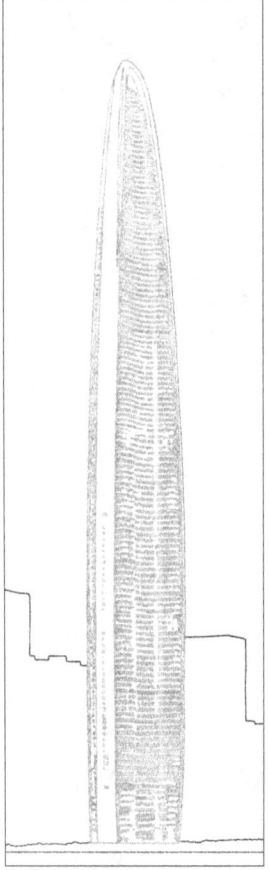

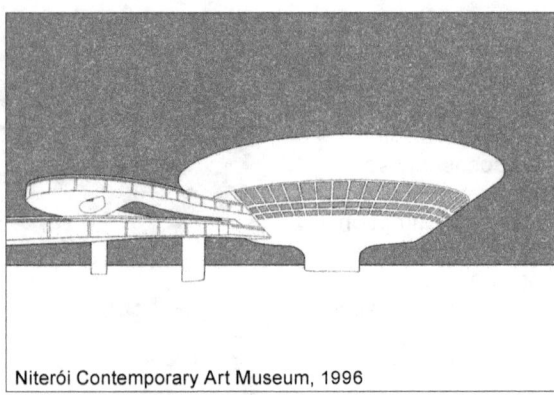

Niterói Contemporary Art Museum, 1996

The Brazilian architect Oscar Niemeyer created designs that appear futuristic because of their aerodynamic curves.

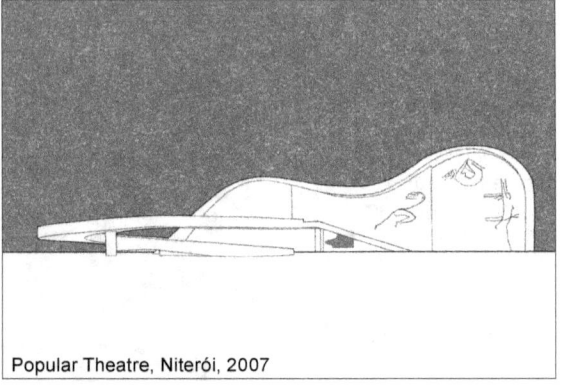

Popular Theatre, Niterói, 2007

On page 75 a section through a cathedral indicates lines of force passing through the structure due to gravity and wind load. When buildings become much higher wind becomes an even greater factor in design. Contemporary skyscrapers, such as the Greenland Tower in Wuhan, China, by Adrian Smith and Gordon Gill (2019) have aerodynamic curves to mitigate wind forces, but also to make them beautiful.

Some look like flying saucers, others rippling waves, and his museum (below) has a cross-section like an aeroplane's wing.

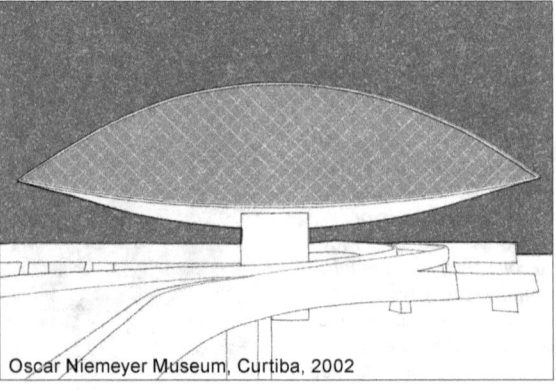

Oscar Niemeyer Museum, Curtiba, 2002

TURBULENCE
Winton Gallery, Zaha Hadid, 2016

Expressed in numbers, the fascination of mathematics is less visual than intellectual. When the London Science Museum's Winton Gallery – devoted to the contribution of mathematics to the world – was to be redesigned, the architects looked for ways of giving the gallery a visual dynamic that numbers could not. They took as their starting point a 1929 Handley Page biplane and modelled the turbulence it would have caused as it flew through the air. This provided them with a dramatically curvaceous sculptural form that they could use to give the gallery an arresting identity.

The turbulence behind an aircraft creates complex swirls and eddies. Given visible form with the help of computer software these can be re-imagined in sculptural form.

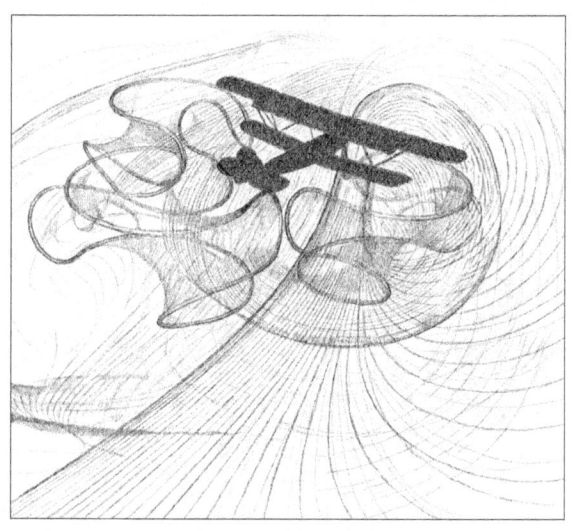

The final sculpture of turbulence in the Winton Gallery is formed of a curved metal framework across which tailored fabric is stretched (below). The tension in the fabric creates its own curves that contribute to the aesthetic power of the piece.

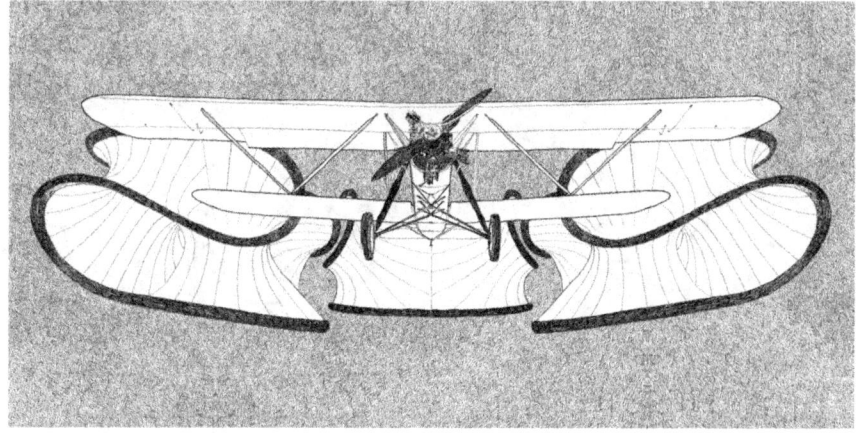

VORTEX
Maggie's Centre, Swansea, Kisho Kurokawa, 2011

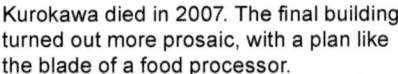

spiral galaxy

Vortexes can be vast or tiny. Kurokawa drew a vortex as the idea for a Maggie's Centre in Swansea (below right).

Kurokawa died in 2007. The final building turned out more prosaic, with a plan like the blade of a food processor.

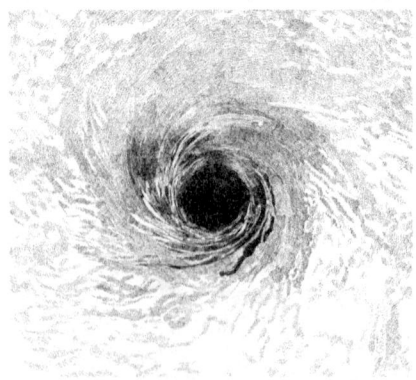

plughole vortex

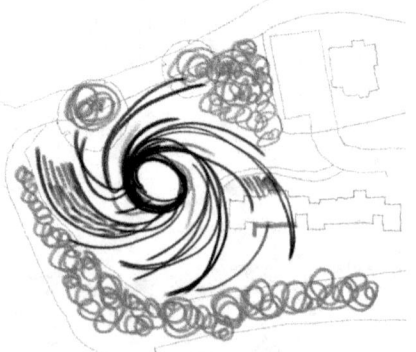

my impression of Kurokawa's sketch

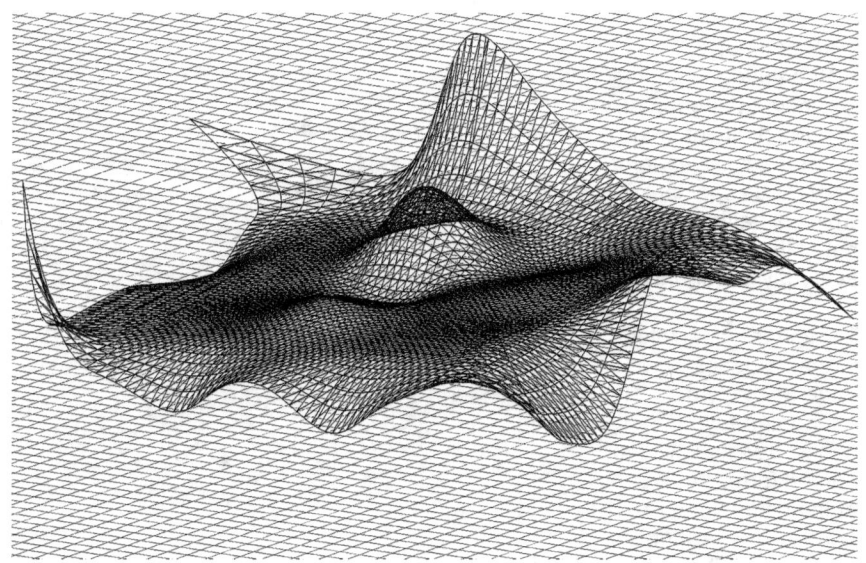

ORCHESTRATING CURVES

In the twenty-first century we have sophisticated ways of manipulating curves. But the evidence of architecture since earliest times suggests we have been fascinated with the construction and orchestration of curves through almost all history. It is easy to construct what we might think of as random curves, merely by gesture or moving our arms. But what we seem to aspire to most is the ability to control curves. We appreciate instruments for the construction of curves almost as if they allow us to perform magic. Because of this, places – architecture – constructed with controlled curves attain special value. Sometimes such value is associated with religious sacredness, sometimes with personal standing. The curve may be expensive and onerous to produce in architecture but it also confers, and is interpreted as an accoutrement of, power and status.

ORCHESTRATING CURVES IN PREHISTORY
the circle

Ideal geometry, mathematical geometry, the geometry of perfect circles and squares, exists in some mysterious special zone which is neither of nor not of the world of our everyday experience.

Though the principles of mathematics may underlie many natural phenomena, you cannot see it explicit in natural form. And yet, at the same time, its universal dependability and applicability means that it cannot be a pure creation of the mind. It belongs to a special place.

There are infinite irregular loops by which the boundaries of a place might be defined. But, though it may vary in radius, there is only one perfect circle. The regular curve of a perfect circle is a representation of right. Other loops are deviations from this emblem of incontrovertible virtue.

When we invented mechanisms by which we were able to introduce ideal geometry into the world it must have seemed as if we had acquired magic powers. It is no wonder that a stick could become a magic wand, an instrument of power by which perfection could be made manifest.

There are any number of irregular loops that can identify and define the edges of a place. By hand and eye we cannot ensure that a circle is geometrically perfect. To do that we have to use mechanical means (below). The pin and string act like gravity, holding the pencil in a perfectly circular orbit around the centre.

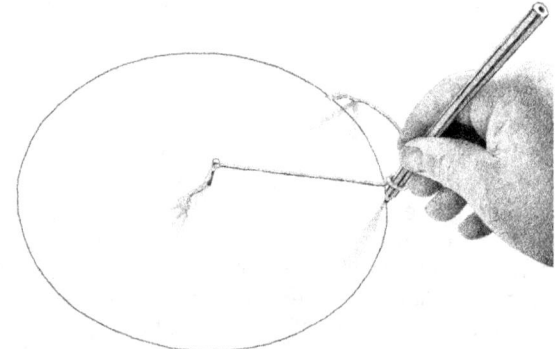

Using this simple device the curves of a circle are controlled and regular, free of the vagaries of imperfection to which our own eye and hand are prone. For any particular radius, there is only one perfect circle. A circle, is a circle, is a circle… It brings with it something unworldly into the world. Rather than 'any old loop' defining a place, the circle is the one and only 'right' loop, and therefore possesses and disposes a qualitative moral status of its own, as befits important ceremonial sites.

ORCHESTRATING CURVES IN PREHISTORY
the first mechanical computer

There is a qualitative difference between a rough loop of stones and a perfect circle (right). The difference is achieved by the use of a mechanical device – a 'computer' – by which the inaccuracies of hand and eye can be obviated and transcended.

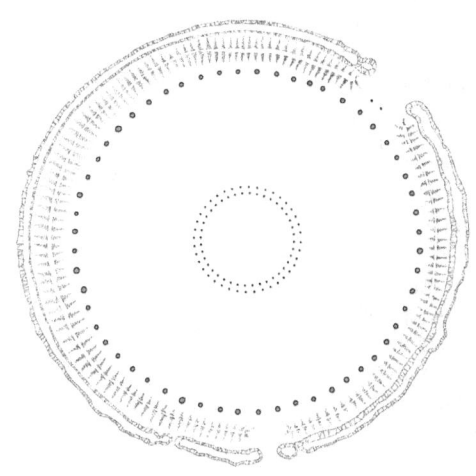

early Stonehenge circles

The stones of early versions of Stonehenge were laid in perfect circles presumably achieved by the employment of that early 'computer' comprising a stake in the ground and a length of rope.

Twiddling a pair of compasses is fascinating even now in the twenty-first century. Many school exercise books across the world are decorated with 'flowers' constructed by drawing circles in a circle (above). But imagine the fascination in the mind of the person who first realised that by sticking a peg in the ground a perfect circle could be circumscribed using a length of hide or gut. And with ropes, the circles could be bigger. Only perfect circles could be good enough for high status ceremonial sites.

Apple HQ, Cupertino, CA

And now, thousands of years later, the product of that first computer – the perfect circle – is considered the most apt form for the headquarters of one of the pre-eminent manufacturers of modern computers.

ORCHESTRATING CURVES IN PREHISTORY
Woodhenge

In prehistoric times it seems we also developed hybrid instruments for the construction of curves other than the perfect circle. It would have only taken a few experiments in the dirt to develop a mechanical computer (right) for the construction of an ellipse – a more subtle geometric form which, while maintaining regular curves, has two foci or centres rather than one.

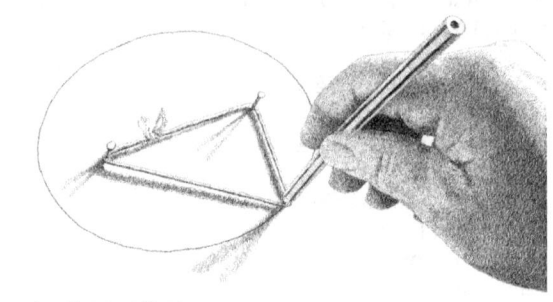

constructing an ellipse

Whereas the circle has just one centre, and which therefore architecturally can accommodate only a single dominant entity – an altar or a priest – the ellipse frames a duality, the face-to-face juxtaposition of, for example, a worshipper with the focus of his or her worship. It is an arrangement in which both are acknowledged architecturally.

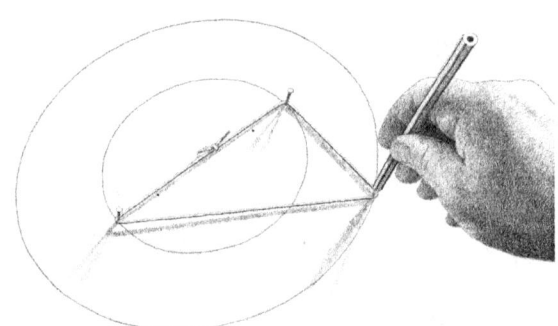

constructing concentric ellipses

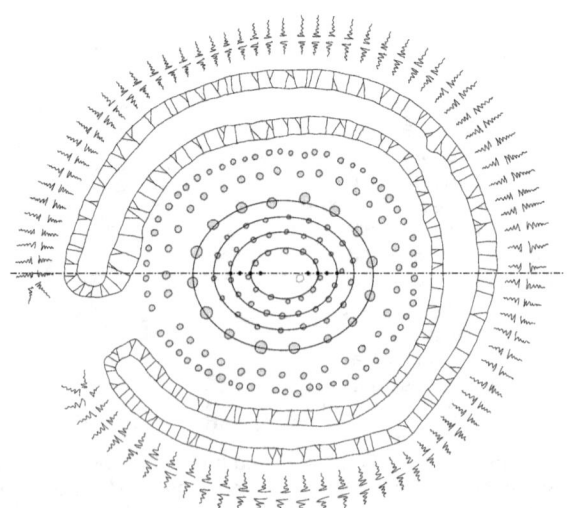

Ellipses are evident in ancient sites such as Woodhenge (right) where concentric ellipses of tree trunks were erected around a sacred place with two foci.

plan of Woodhenge

ORCHESTRATING CURVES IN PREHISTORY
later Stonehenge

The ellipse is evident too in the later versions of Stonehenge (right).

Stonehenge represents one of the earliest sophisticated orchestrations of a range of curves in architectural history. It comprises a pair of non-concentric ellipses and a number of concentric circles. These curves accomplish a composition that combines identification of place, relation to axis – that of the sun setting on the winter solstice together with a line of approach from the north east – and an elliptical duality in which the priest or worshipper can position him or herself in juxtaposition with a fixed focus.

In these ways, without the distraction of corners, an orchestration of curves realised in standing stones and trilithons frames and makes sense of ceremonial events and relationships. This is architecture.

For alternative theories on the geometric construction of Stonehenge see:
Anthony Johnson – *Solving Stonehenge*, 2008.

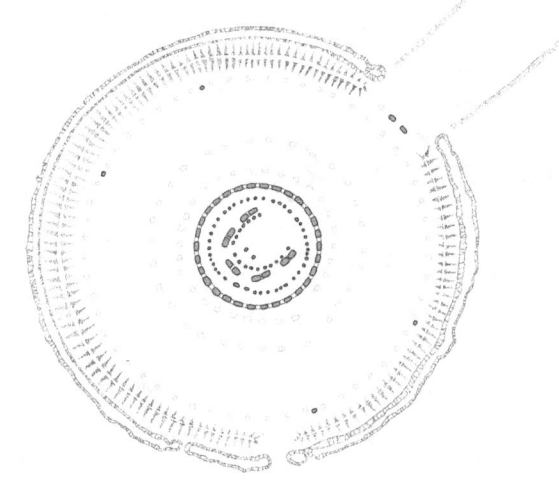

the late plan of Stonehenge

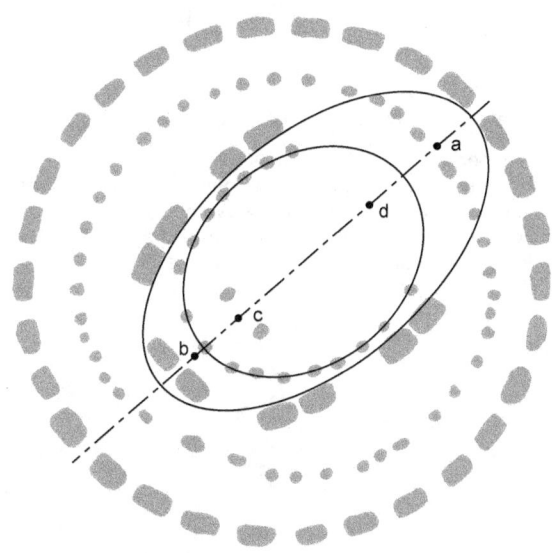

The inner stones of Stonehenge (above) are arranged in a composition of ellipses. The outer horseshoe of trilithons is arranged on an ellipse with foci near the entrance (a) and between the main trilithon (b), which is set inside rather than on the perimeter of the ellipse. The inner elliptical horseshoe of smaller stones has foci at c – the focus of the whole monument, where there may have been an altar or a ceremonial portal – and at d, where a principal priest or important participant in the ceremonials would probably have stood.

ORCHESTRATING CURVES IN ROMAN TIMES
Hadrian's Villa at Tivoli

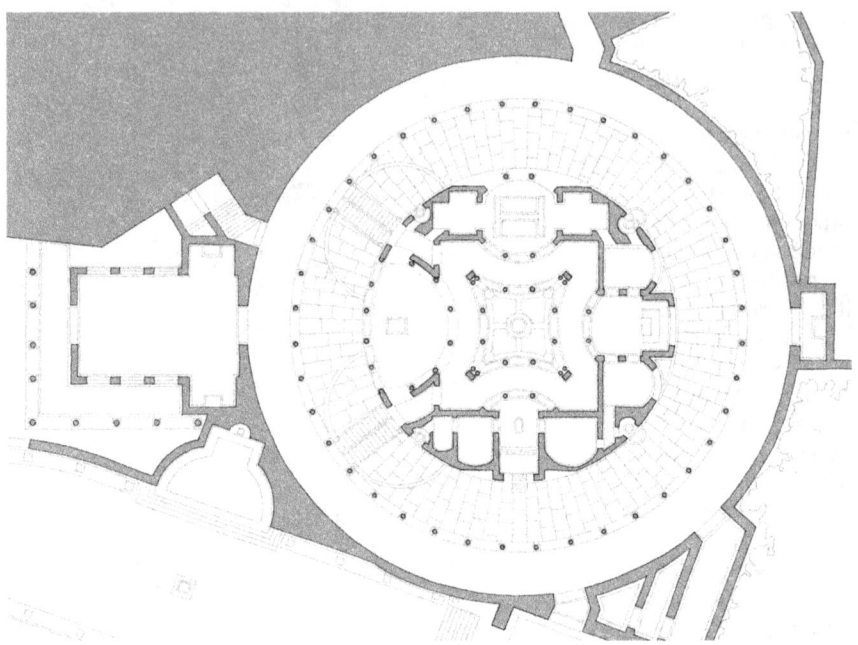

Circular Pavilion or Maritime Theatre

The imperial status of Hadrian's Villa is displayed in the employment of complex curved compositions for the plans of various of its pavilions and other buildings – e.g. the Circular Pavilion (above) and the Women's Baths (below). Curves – more difficult to lay out, more expensive to build – are, as they continued to be in the twenty centuries following, accoutrements of power and expressions of wealth and sophistication.

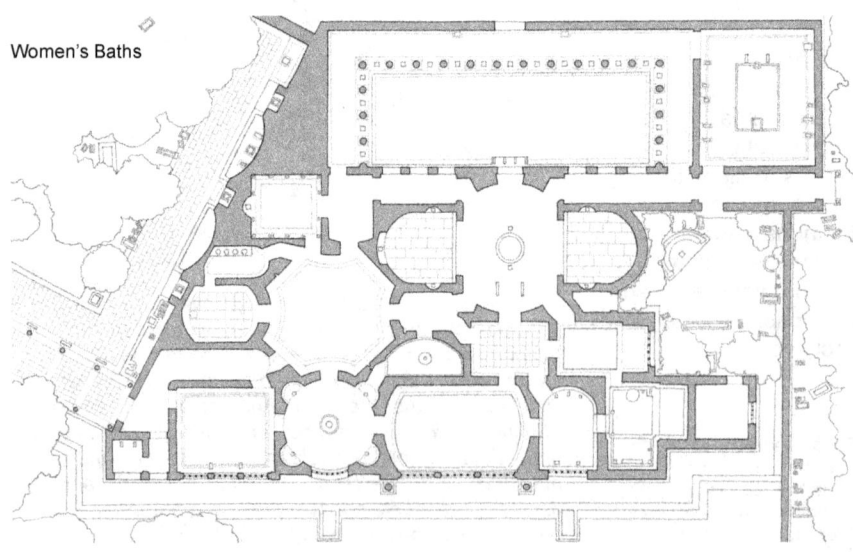

Women's Baths

BYZANTINE CURVES
Hagia Sophia, Istanbul, 6thC

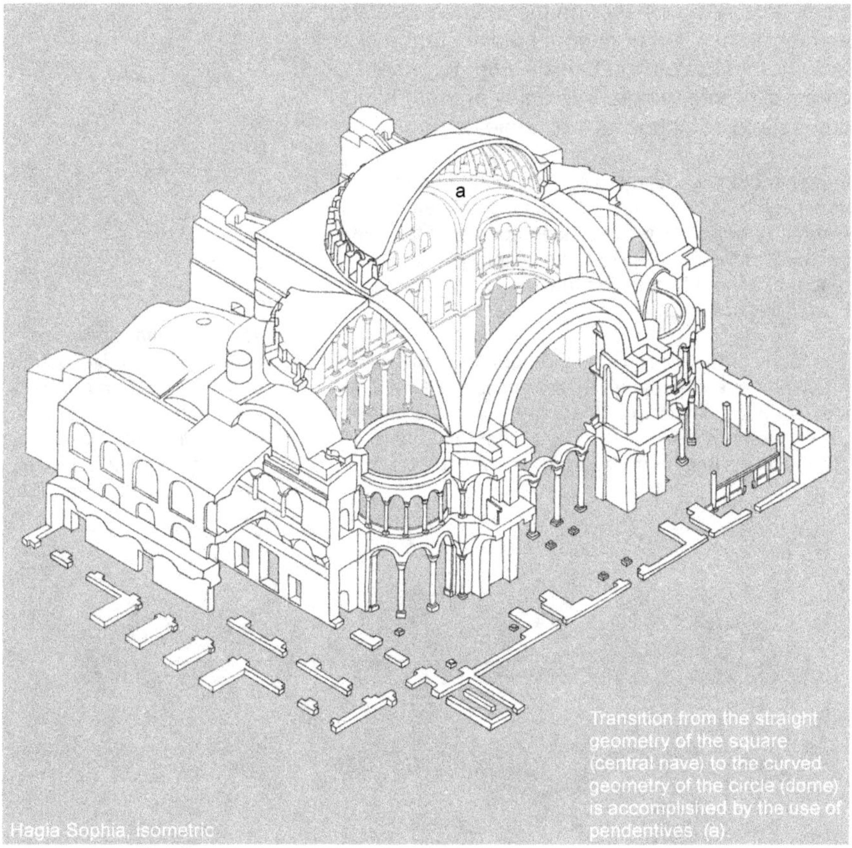

Transition from the straight geometry of the square (central nave) to the curved geometry of the circle (dome) is accomplished by the use of pendentives (a).

Hagia Sophia, isometric

Hagia Sophia (above and right) was built in the sixth century CE. Although its great dome and supporting half-domes have occasionally partially collapsed due to earthquakes, the building's present form has survived, more-or-less, for nearly a millennium and a half. Hagia Sophia has been a church and a mosque, and is now a museum. Its symphony of curves is a hymn to the religious soul but also to the human intellect and the skill with which it was built.

Hagia Sophia is one of the most symphonic orchestrations of curves in all architecture. All the curves are in some way structural but also symbolise paradise. It is a hierarchy of curves from the human to the celestial.

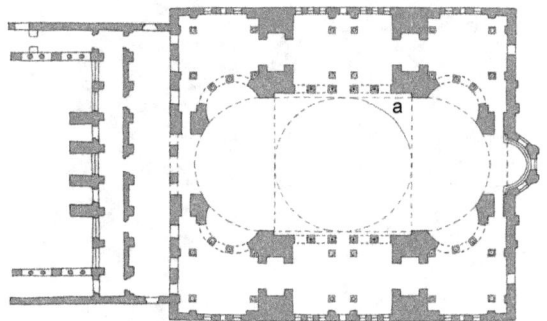

Hagia Sophia, plan

CURVE

149

ARCADE AND MIHRAB
Mezquita de Córdoba, Spain, 961 CE

Sometimes it is not enough that a curve is a curve, any more than it is enough, in music, that a chord is a chord. Curves and chords work together to create a greater whole. We enjoy orchestrating curves into complex but harmonic compositions.

In architecture we take the orchestration of curves seriously. It has been so for thousands of years. Geometry helps decision-making. It holds a fascination because of the congruences and alignments, resonances and harmonies that can be created visually in architectural form.

The focus of the Mezquita in Córdoba is its mihrab (opposite; see also page 69), a scintillating arch indicating the direction of Mecca. Around it is an arcade of complex arches (below). The arches and the mihrab are, as illustrated, complex matrices of circles and squares.

The visual harmonics of the arcade is achieved not only by repetition but by a complex construction of geometrical figures including circles, squares and an equilateral triangle.

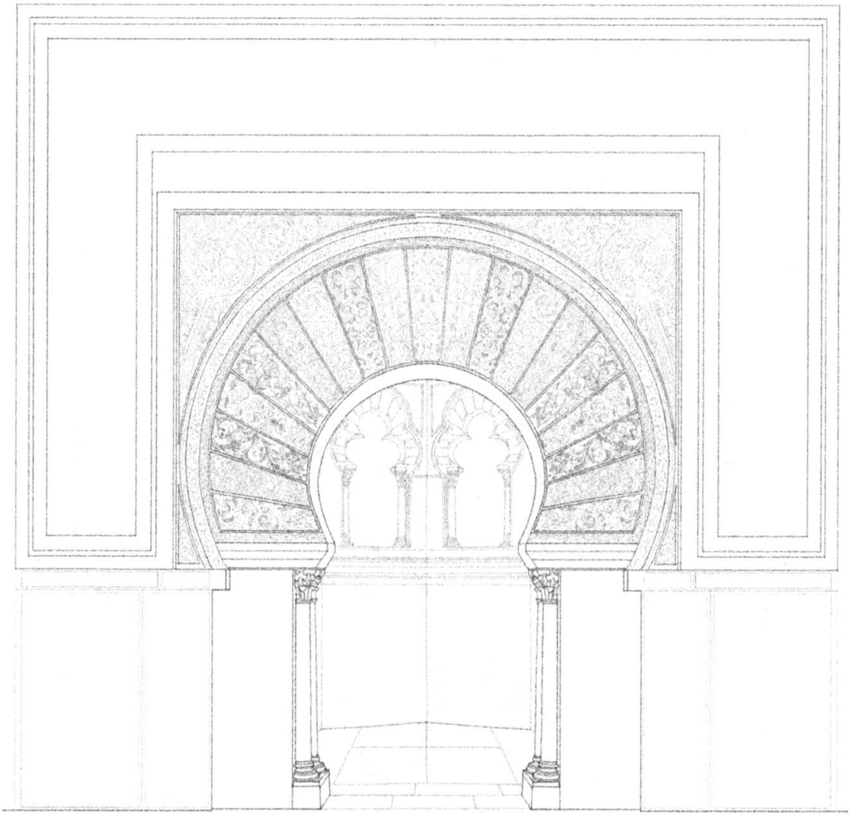

mihrab of the Mezquita in Córdoba

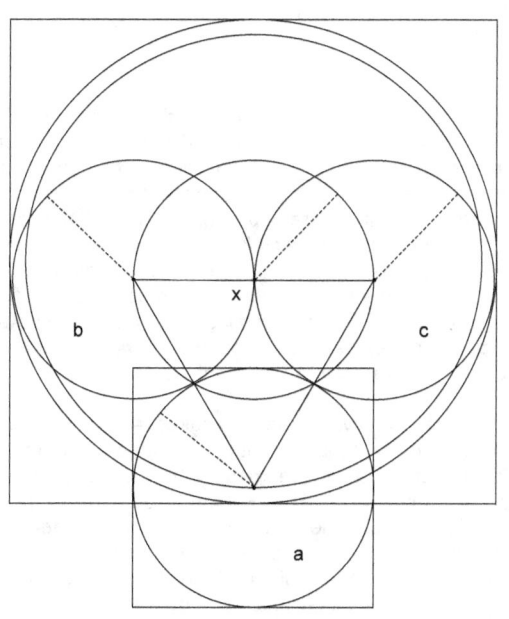

The Córdoba mihrab is decorated with intricate designs based on the curving forms of plants. The radiating lines make the opening shine like the sun. The mihrab's own curves are based on a simple but subtle composition of circles and squares (right). The design starts with the smaller circle in a square (a). Two more same-sized circles (b and c) touch it and each other, making an inverted equilateral triangle. Where the two upper circles touch gives the centre of the circular opening (x). The larger circle in a square just touches the outer perimeters of the two upper circles.

CURVE

ROMANESQUE CURVES
Notre-Dame d'Avy, Charente Maritime, 12thC

This is a doorway into a church near the west coast of France. Originally Romanesque, a distinguishing feature is the ornamented round arch consisting of a number of concentric arcs exaggerating the perspective depth of the entrance. Curves are apparent not only in the arcs of the layers of arch but also in the twisted columns, the carvings of birds and beasts, and some intricate knot-like relief work. Some stones seem to have been reused from earlier buildings. The mason's fascination with the power of a pair of compasses is evident in ornamental patterns such as that illustrated on the right (at a above).

The church was given a Gothic 'makeover' in the fifteenth century, as is evident in the shape of the actual door opening in the above image, which has been given a pointed (as in the Al-Aqsa Mosque, see page 74) rather than a round arch. The curves of a pointed arch seem more elegant as well as being appropriate to Christian architecture in that they differentiate the arch from the Roman round arch; they also point heavenward.

STRUCTURAL CURVES INSPIRED BY TREES
in Gothic architecture

As can be seen in the *Metaphor* Notebook (a companion to the present volume) trees can hold metaphorical power over architecture (see pages 45 and 47 of that other Notebook, from which the images on this page are repeated). In Gothic architecture trees influence the structural aesthetics of buildings too.

The converging curves of the vaulting of great cathedrals are reminiscent of the converging boughs of great trees in an avenue (below, top pair). The spreading canopy of the vault of a chapter house is like the spreading canopy of a single tree (below, bottom pair).

The vaults of a cathedral ceiling curve together like the boughs of great trees.

They shelter the ceremonial pathway like the canopies of an avenue of trees.

The vault of a chapter house identifies a place in the same way as a single tree.

CURVE

CURVES IN GOTHIC ARCHITECTURE
Wells Cathedral, England, c.12th to c.15thC CE

Gothic cathedrals are multi-layered orchestrations of curves. They constitute the architectural equivalents of great religious oratorios; perhaps they are even greater expressions of the human capacity to imagine and create. In cathedrals the curves are generated according to structural order and for the purposes of ornamentation. The curves are governed by the pair of compasses as a mechanical device for drawing circles. All is disciplined by the authority of the central axis of symmetry and the doorway axis of aspiration that together focus the whole composition on altars positioned on the building's centre line.

The Freemasons' logo (below) is a set square and pair of compasses. The G is for geometry (and God).

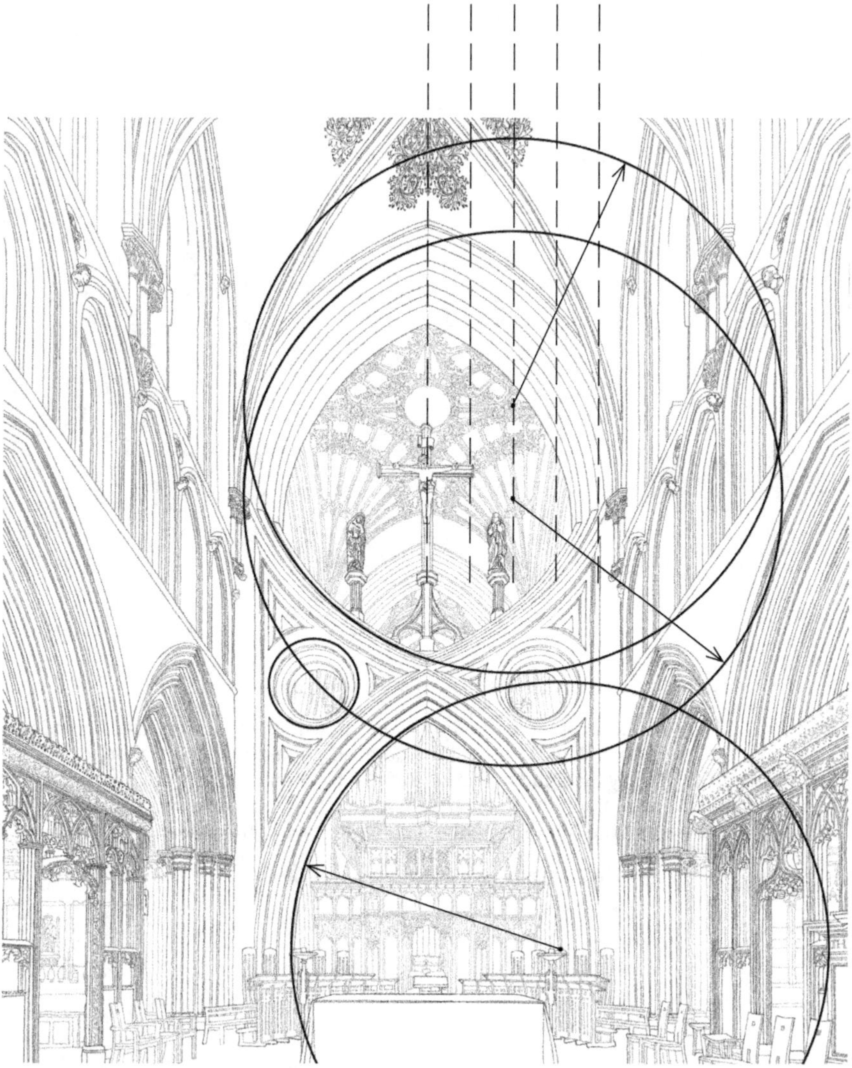

Cathedrals are manifestations of geometry. That geometry is governed by the set square but also by compasses. These governors give shape to every element of the architecture from the grandest structure down to the smallest detail (e.g. the column plans below). The building represents the discipline and beauty of geometry.

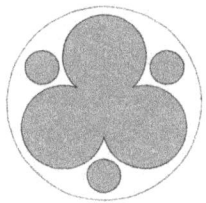 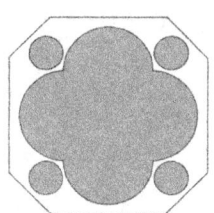 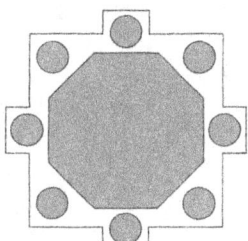

CURVE

CURVES IN GOTHIC ARCHITECTURE
tracing floors

surviving tracing floor at York Minster (after Harvey's 1968 record)

The curves of the elements of the cathedrals were drawn out on plaster 'tracing floors'. This is where the masons would work out the detailed geometry of the tracery of the intricate windows usually included in Gothic architecture, and prepare templates for the stone cutters. These floors illustrate the importance of the discipline of the curve in Gothic architecture and the authority of the set square and compasses.

surviving tracing floor at Bourges Cathedral, France (thirteenth century)

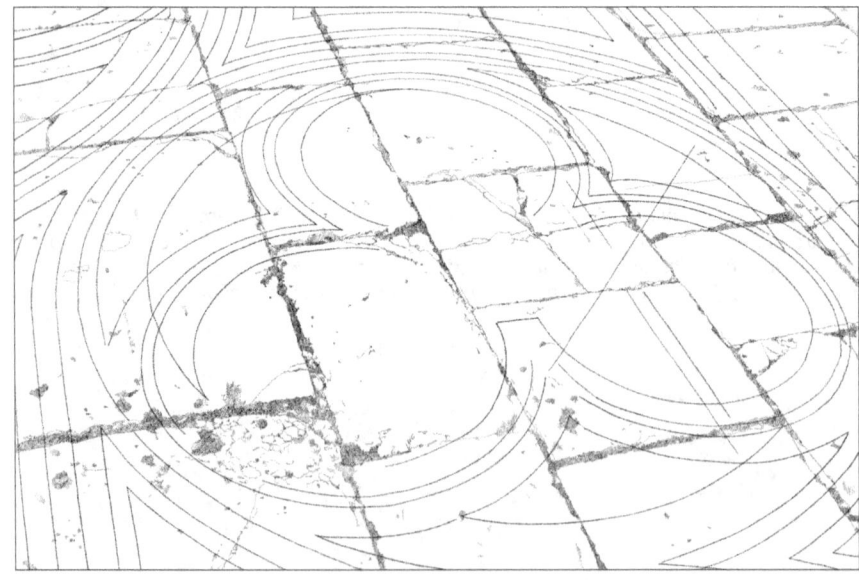

CURVES IN GOTHIC ARCHITECTURE
floor and paper as computer screen

The lines set in the rectangle of the tracing floor of a loft in York Minster (right*) are like drawings on a medieval computer screen. The plaster surface establishes an otherworldly realm on which complex curvaceous geometries can be worked out. The mechanical processors of this computer, divined by the mind of the architect, are the pair of compasses and the straight edge.

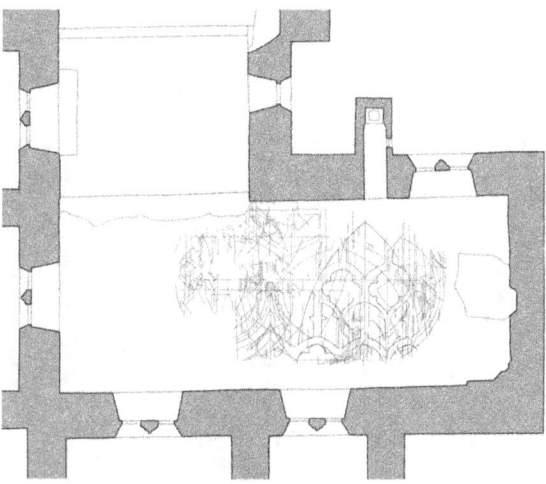

The Platonic ideal geometry that governed the form of the resulting building carved in stone persists as a palimpsest, a ghostly image on the tracing floor.

The masons who used tracing floors to draw out their designs at full scale probably also used paper to develop their architecture at smaller scales. One of the earliest surviving medieval drawings (left) dates from the thirteenth century. It depicts part of Strasbourg Cathedral and is entirely constructed of straight lines, right angles and curves governed by a pair of compasses. These constitute the authority underlying the architecture.

* See Holton – 'The Working Space of the Medieval Master Mason: the Tracing Houses of York Minster and Wells Cathedral', 2006.

CURVES IN GOTHIC ARCHITECTURE
trefoils and quatrefoils

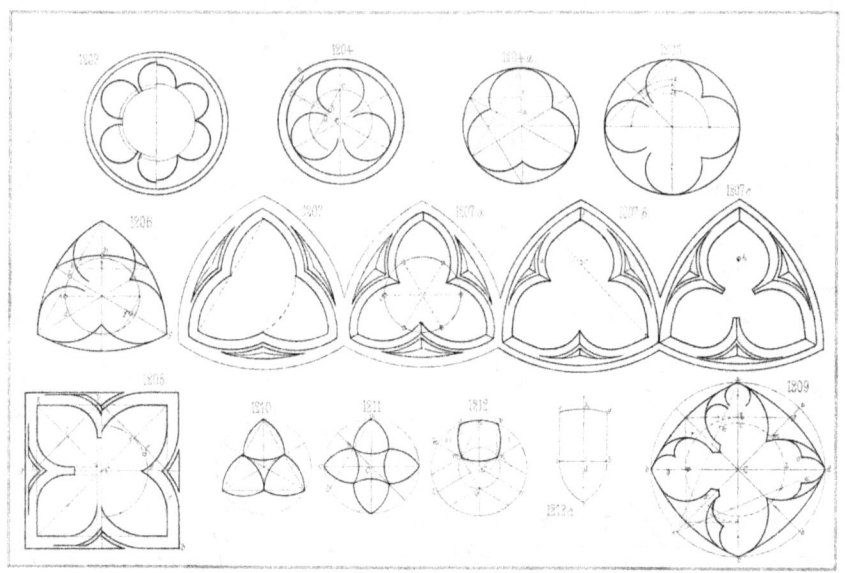

Ungewitter – *Lehrbuch Der Gotischen Konstruktionen* (1858), 1901.

Pre-modern (mainly nineteenth-century) textbooks for architects across Europe included advice on the construction of Gothic ornament and tracery.

All is governed by the transcendental rule of geometric curves.

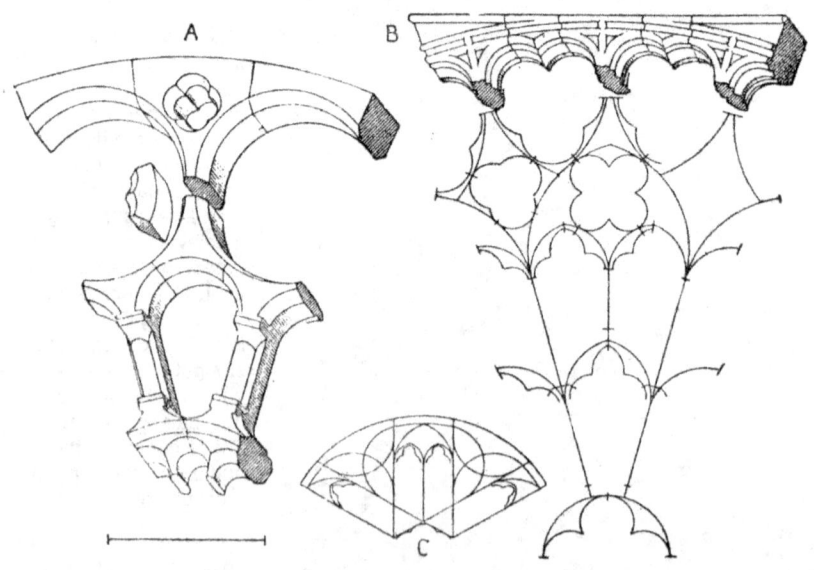

Choisy – *Histoire de l'architecture* (1899), Volume 2, 1964.

CURVES IN GOTHIC ARCHITECTURE
tracery

Meyer – *Handbook of Ornament* (1894), 1987.

CURVES IN GOTHIC ARCHITECTURE
rose window, Notre Dame, Paris, 1260

The geometry of Gothic tracery was constructed in an elaborate version of the game school pupils play with their compasses, twiddling patterns on their exercise books (see page 145 and Meyer's *Handbook of Ornament* on page 159). But in the case of the great windows of cathedrals this process was invested with a belief that it was drawing on the generative methodology of the universe; i.e. that it was the approach employed by God the Great Architect.

The thirteenth-century rose window (above) from the north transept of Notre Dame in Paris is geometrically constructed according to an intricate and sophisticated version of a doodle using a pair of compasses.

In the 1790s the English poet and illustrator William Blake depicted Urizen (left) – architect of the universe – giving form to the human conception of the world by means of a pair of stonemason's compasses. For Blake Urizen was not the 'true' God but a creature of the human imagination.

CURVES IN GOTHIC ARCHITECTURE
King's College Chapel, Cambridge, 1550

The fan-vaulted ceiling (above) of the chapel of King's College in Cambridge (see also page 69) is possibly the most astonishing manifestation of Gothic architects' and builders' ability to manage curves in three dimensions. It marked the culmination of Gothic ecclesiastical architecture, which was about to be superseded by Renaissance classicism – an architecture ruled by Platonic geometry.

Without the help of a computer few people in the twenty-first century would have the first clue how to determine the exact three-dimensional curves of this ceiling's components.

A MAJOR CHORD...
Villa Rotonda, Vicenza, Andrea Palladio, 15thC

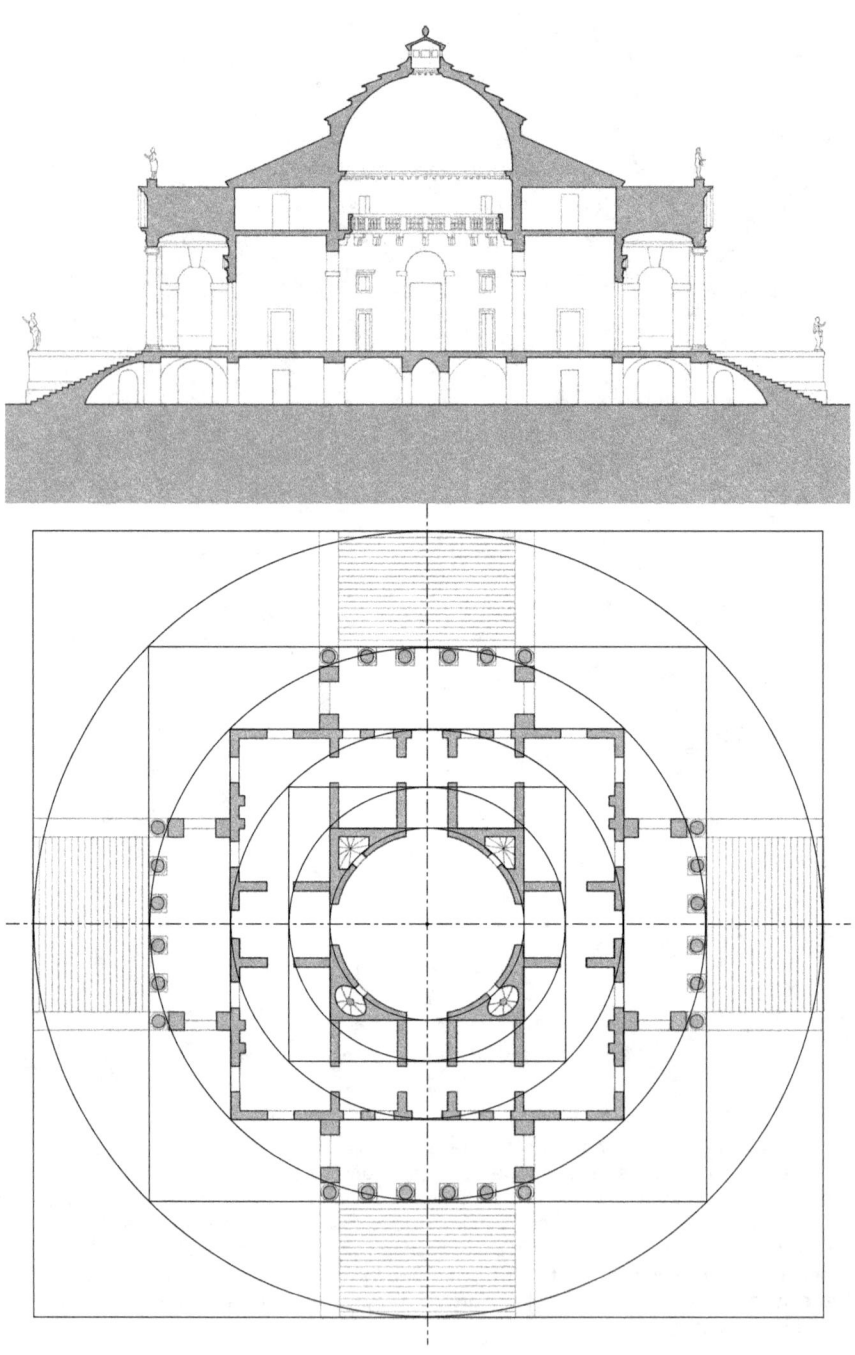

... AND A HYMN TO St PETER
Il Tempietto, Rome, Donato Bramante, 1500

In Italy during the latter part of the fifteenth century, architects sought in their designs the visual equivalent of harmony in music. In this search the curves of the circle described by a pair of compasses played equal partner with the straight lines and right angles of the square.

Palladio's Villa Rotonda (opposite) was constructed on a geometric framework of concentric circles and squares that spread from the centre like ripples from a stone dropped into a pond.

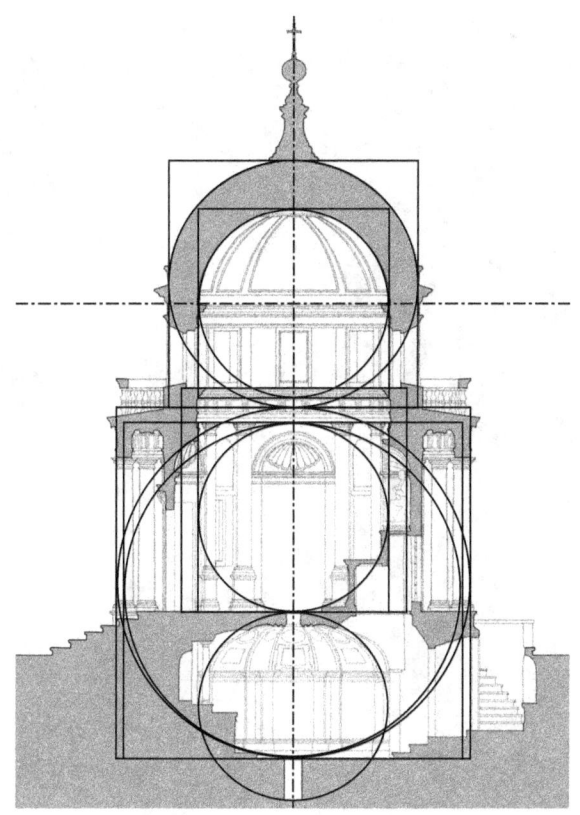

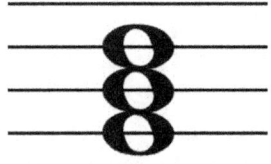

Bramante's Il Tempietto (right) is a subtle composition of overlapping circles (hemispheres and cylinders) and squares that ring visually and spatially like the notes in a musical chord.

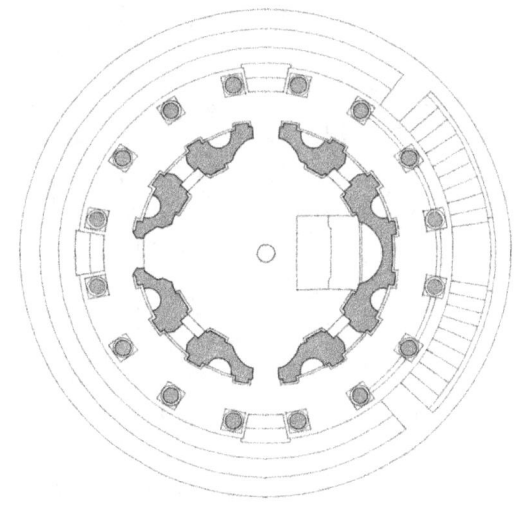

CURVE

TAJ MAHAL
Agra, India, 1643 CE

It is often tricky to identify the exact geometry underlying a work of architecture.

Sometimes there is none. But in the case of the Taj Mahal geometry does appear to be there.

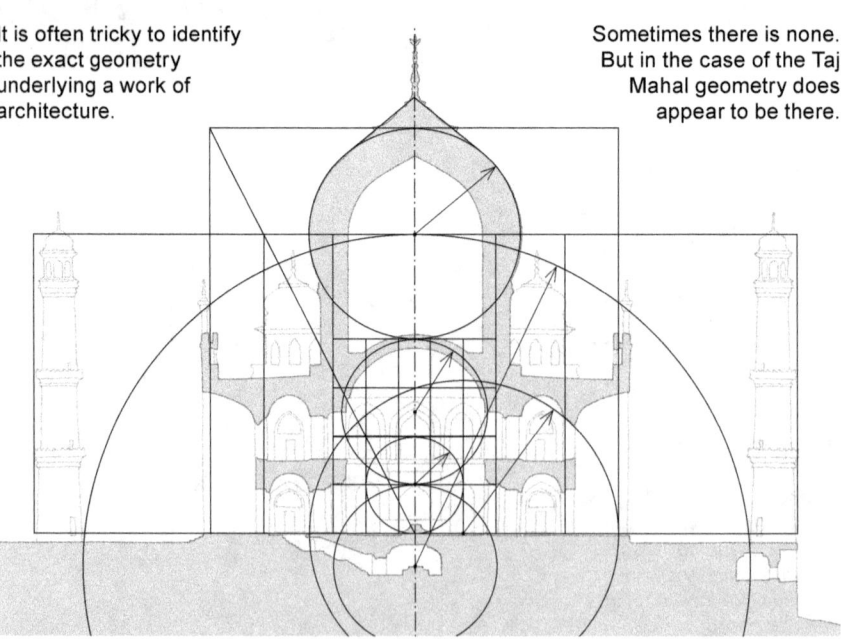

BAROQUE-ROCOCO CURVES
Vierzehnheiligen, Bavaria, Balthasar Neumann, 1772

Some of the most extravagant use of curves in architecture occurred in European Baroque and Rococo buildings of the eighteenth century. The aim in these cases seems to have been less to realise in material form some divine geometry than to obfuscate space and dematerialise fabric. The experiential effect of the elliptical geometry of the plan of Vierzehnheiligen Church (below) is to leave the visitor not quite knowing where the boundaries of the interior actually are.

Every straight line is broken up with curved ornamentation. This further aids the apparent dematerialisation of the fabric of the building. The plan is composed of circles and ellipses (below).

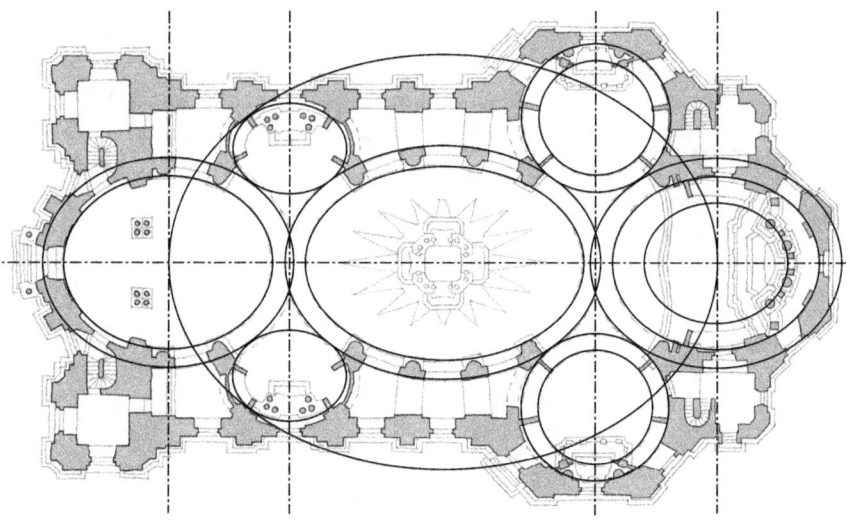

ÉCOLE DES BEAUX-ARTS, PARIS
J.-A. Leveil, Architecte, 1880s

Towards the end of the nineteenth century the École des Beaux-Arts in Paris was the pre-eminent school of architecture in the world. It promulgated a form of Classicism based on refined versions of Graeco-Roman precedents. The school's tutors were intent on codifying the principles and details of beautiful architecture.

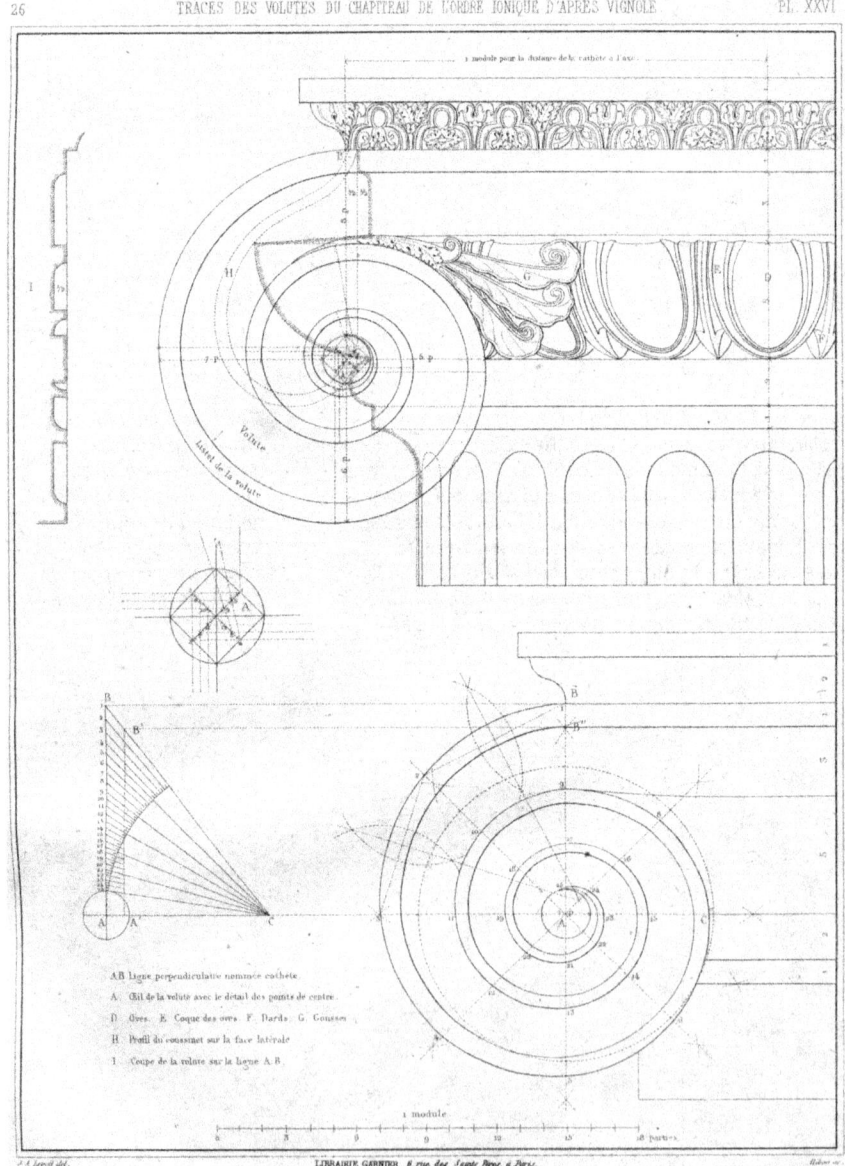

It was as if the École des Beaux–Arts was intent in making the art of architecture into a science by which predictable results could be achieved by following precise methodologies for design. For instance, the drawings on these two pages are taken from a manual of design by J.-A. Leveil produced in the 1880s. They give the student architect precise directions for the construction and orchestration of curves to produce an Ionic capital. Note that the volute spiral is produced using compasses rather than a whelk shell (page 112).

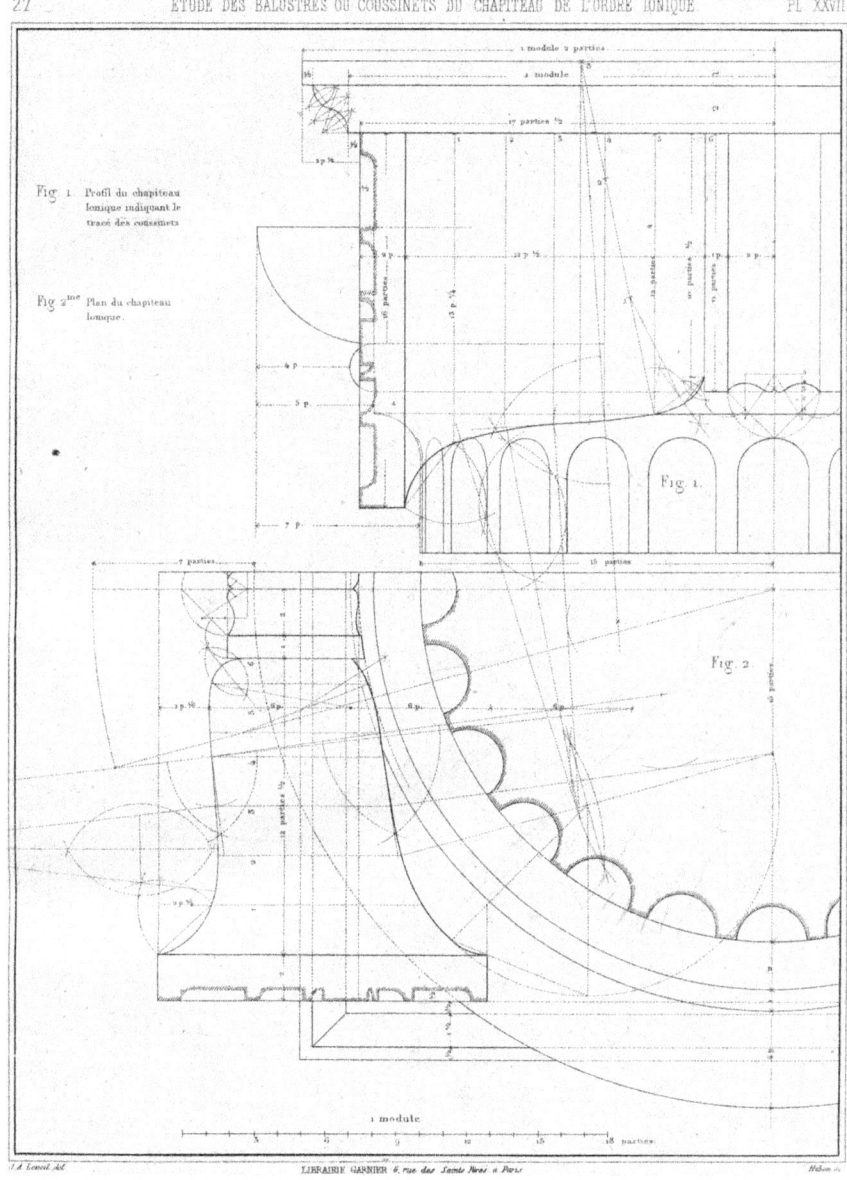

CURVE

Leveil also gave detailed and precise instructions for the design of various types of columns. In the drawing below he sets out: how a Tuscan column should diminish in girth as it rises; the precise curves of a barley-sugar twist or Solomonic column (colonne torse); and the rules for the entasis of a Corinthian column (corinthien renflé). Notice how Leveil's method is different from that described on page 121. Leveil's method was based on the prescriptions of sixteenth-century Italian architect Giacomo Barozzi da Vignola. Authority was sought from ancient precedent. But no incontrovertible geometric rules have been established.

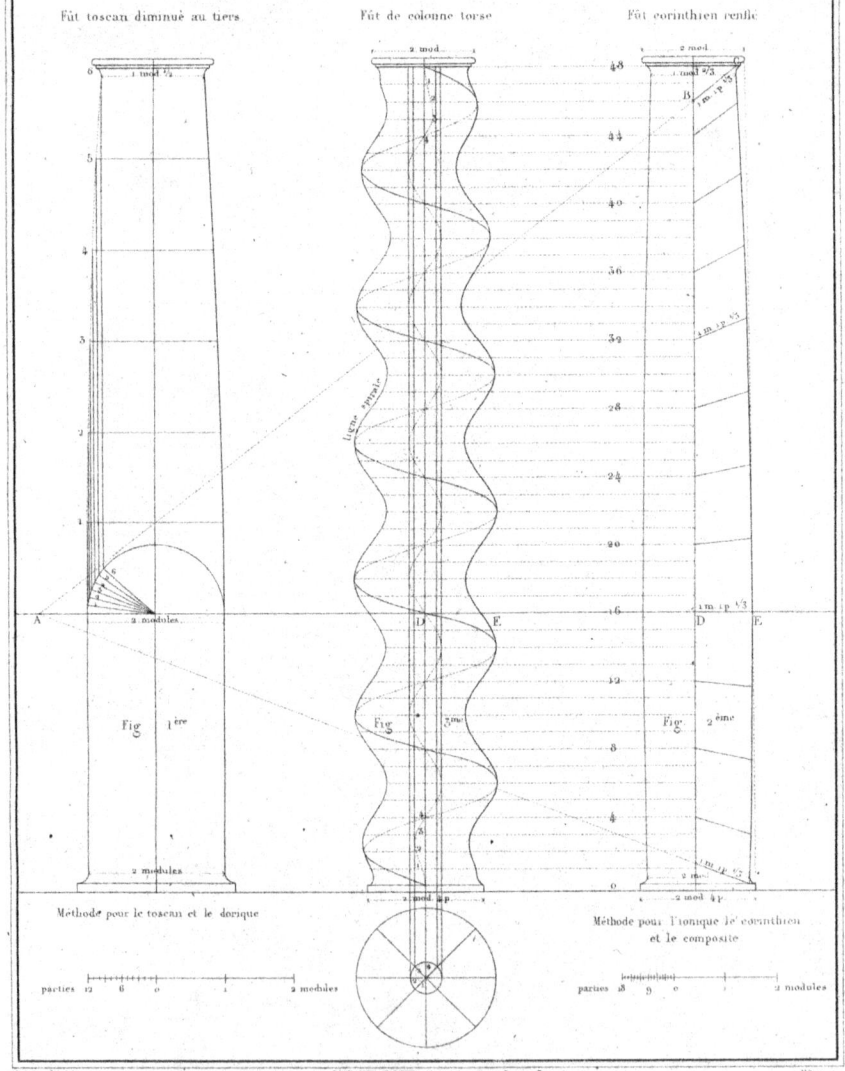

HIDDEN ORCHESTRATION OF CURVES
Villa Mairea, Finland, Alvar Aalto, 1939

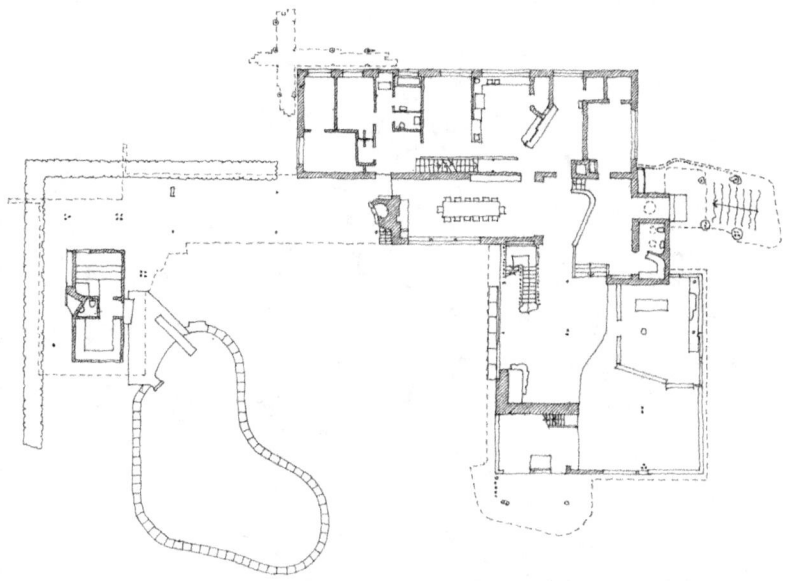

The orchestration of curves in architecture morphs through history. In the twentieth century the explicit use of curves according to specified rules transformed into regimes that were more liberal and individual.

The Finnish architect Alvar Aalto was one who was interested in the role of curves in architecture (see also pages 135–6). His design for the Villa Mairea, although based on the discipline of a grid (below), is also a composition of subtle curves.

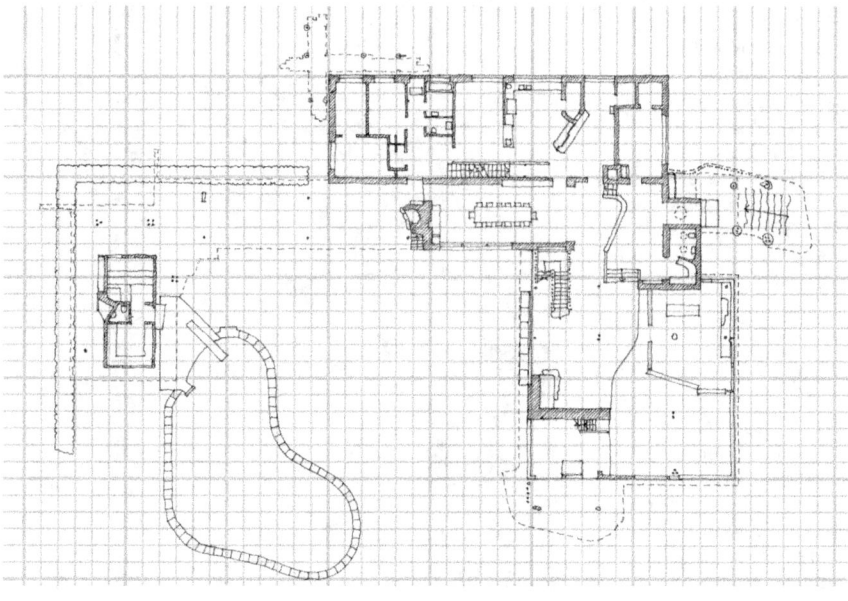

COUNTERPOINT – CURVES AND ORTHOGONALITY
Villa Mairea, Alvar Aalto, 1939

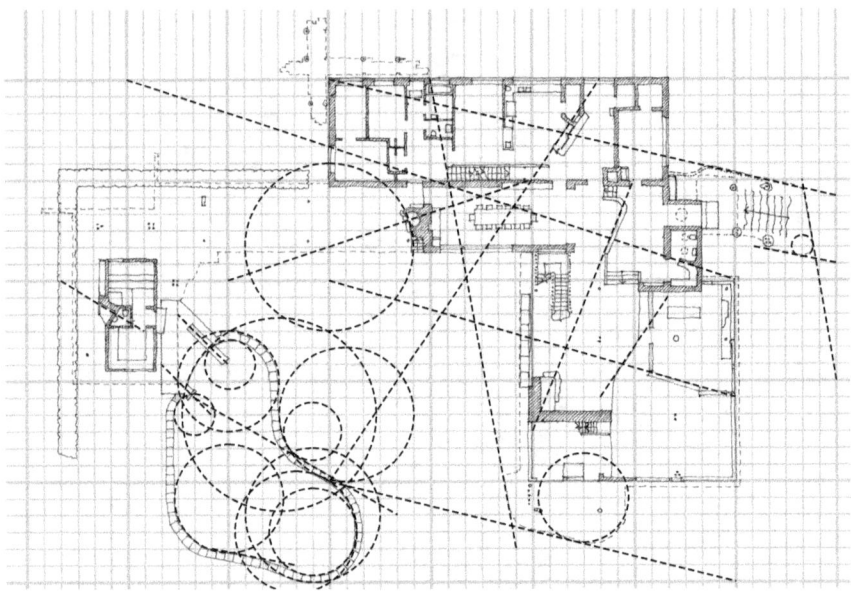

A tentative analysis of Aalto's plan for the Villa Mairea suggests that the orchestration of curves was carefully managed by relation to the underlying grid (above). (In this he may have used french curves.) It is clear that Aalto was interested not in curves as the dominant discipline but as a counterpoint to or fusion with the orthogonal (see page 32). This was his way of expressing the melodic element of architecture.

A deep curved sculptural niche in the rectangular chimney breast of the Villa Mairea (right) illustrates Aalto's interest in melodic–rhythmic/curved–orthogonal counterpoints in architecture. It also illustrates how curves catch light adding sculptural interest.

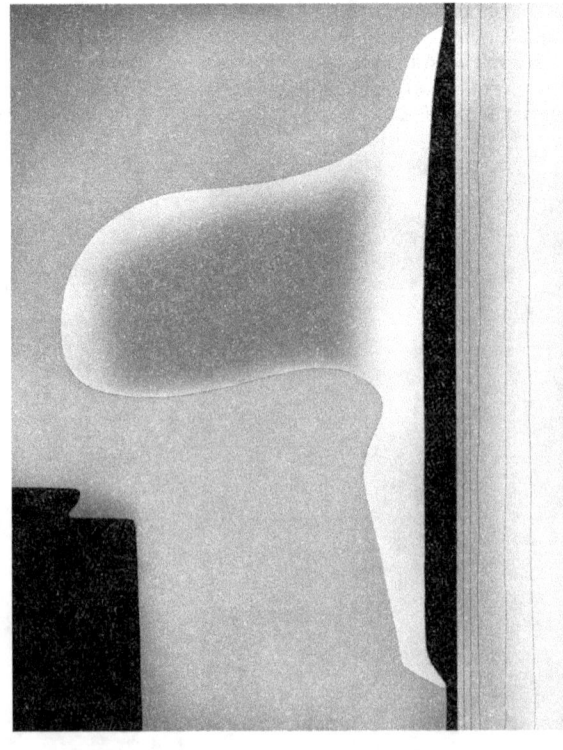

ENDNOTE

Curves? They season architecture but are they good or bad? Do they release ills upon the world or enrich it with vitality and sexiness? Should we be wary of them or embrace them? Should we celebrate their meandering or reject them in favour of sensible straight lines and rational rectangles? Are curves subversive or expressions of the human imagination's ability to transcend the ordinary? Are they symptoms of virtue or turpitude?

The trajectory of the curve in architecture, through thousands of years, is one of adventure: defiance of the mundane combined with determination to achieve the hitherto seemingly impossible; to roof great spaces, to astonish with grand sculptural pirouettes. The adventure is a striving for transcendence. Curves – circles, ellipses – bring realms of ritual and magic into the everyday world. Curves – arches, vaults, domes, suspension cables... – enable us to span greater gaps than we could with straight lines. Curves give us the power to dematerialise the fabric of buildings, to obfuscate space, to conjure atmospheres impossible within cubic rooms. Curves can string out pathways the ends of which we cannot see. Curvaceous forms demand attention, attract wonder and publicity never enjoyed by orthogonal design. The future, it seems, has always been curved.

Curves are associated with claims to imagination, audacity and achievement and (in clients) with status, wealth and enlightened patronage. If you want to be seen to possess the capacity to meet great practical challenges, if you want to be acclaimed as daring and imaginative, if you want to advertise your client's prominence and resources as well as your own genius... design in curves.

With curves we transcend the everyday, we enter a realm of fantasy. But even so, as in many lines of architectural ideology, we still seek authority for our curves. We want controlling methodologies that originate somewhere other than in chance or the whims of our own imaginations.

How should a curve be drawn? How might curves be generated? According to what criteria? For what purpose and with what meaning? We look for answers in different places: in pure mathematics – the realm of unassailable logic; in the precision and control offered by mechanical devices – ropes and pegs, pairs of compasses, templates and computer software…; in the applied calculation of forces and their geometric routing through built structure to the stable ground; in examples set by earlier architects – in Greek, Roman, Islamic… precedent; in examples offered by the often-presumed greatest authority of all, Nature – in shells, pebbles, trees, tendrils, cobwebs, bones, fish, clouds, rippling water, human form, vortices and spiral galaxies… We sometimes even vest authority for curves in the 'free' (mindless) gestures of the 'natural' (uncontrolled) movements of own arms and hands (in the form of squiggles).

So we think that curves can transport us to a transcendent world fantasised by our own imaginations. But we also believe curves enable us to follow nature. We use curves to defy the physical constraints and conditions that are intrinsic to the physical workings of the world. But we also see curved characteristics in nature that we want to emulate. Curves are banners on the battleground against gravity, which conditions columns and walls to be vertical, and the geometry of making, which (by reason of the various factors outlined in the first chapter of this Notebook) suggests straight lines and rectangles.

Curves are emblems of human heroism. But use of curves is also motivated by reverence for nature. We use curves to imitate natural forms – growth, erosion, patterns of movement… In biomimicry we look to nature and the natural behaviours of things for models that inspire, and most often those models involve curves. When we take Nature as our guiding authority we design in curves. But can we design and construct our own creations with the same complexity and nuance as achieved thoughtlessly by the processes of nature or are our aspirations fatally flawed by our own artifice?

In the second decade of the twenty-first century there has been a proliferation of images of curvy projects on Internet

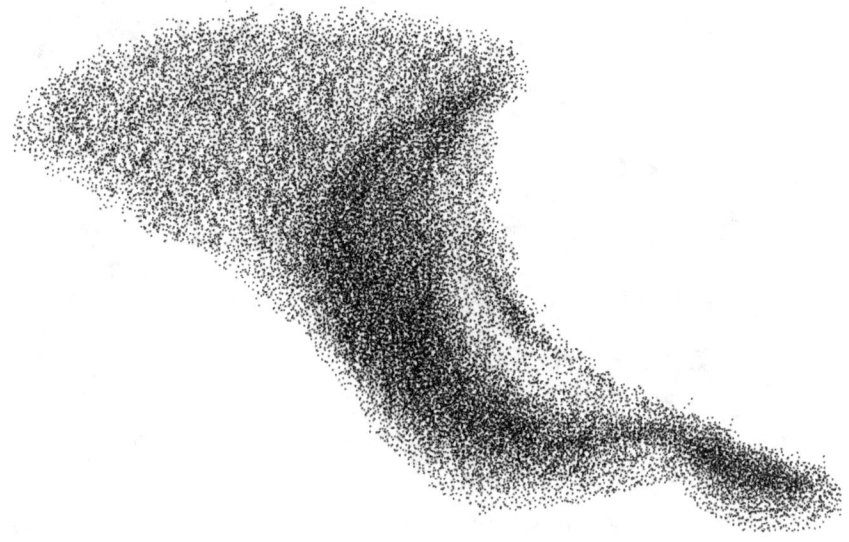

Even a swirling flock of birds can now be recreated by computer-controlled drones. (As in Intel's drone light shows).

See: intel.co.uk/content/www/uk/en/technology-innovation/aerial-technology-light-show.html (April 2018).

architecture news sites – Dezeen, Archdaily... They are evidence of the substantial advances achieved with the help of computer programs, the sophistication and flexibility of which help to free designers and their builders from the restraints of orthogonality – the straight line, the right angle, standardised components... With those advances the subtleties of biomimicry are becoming more and more viable. For example, in the opening ceremony of the Winter Olympic Games in Pyeongchang, Korea 2018, the computer chip firm Intel astonished audiences around the world with a swarm of light-bearing drones directed by computer code to arrange themselves into the Olympic rings. The spectacle suggested that it was possible for human ingenuity to replicate the murmurations of starlings (above) whose synchronised swooping and soaring against the dusk sky have fascinated us probably since prehistoric times.

It seems there is no natural curved form that cannot be replicated with computer software. With 3d printing it seems most of those forms can also be made into models in real space. With computer-led manufacturing processes many curvaceous forms can be realised as actual buildings that prompt awe and admiration. The curve has reinvented architecture as an arm of entertainment.

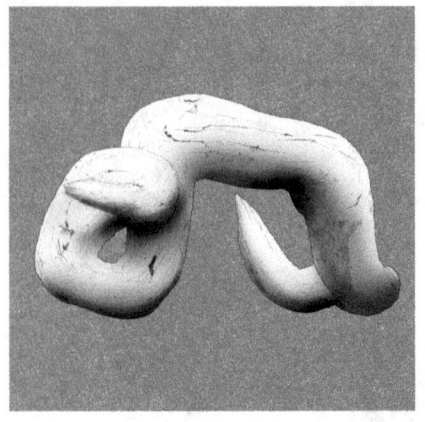

With the aid of computers it is now possible to emulate many complex natural curves.

But curves can be problematic. The curve metaphor implies they can deceive, thwart, confound; they can seduce the unwary into disaster. Domes fall, arches burst apart, bridges vibrate into collapse. Those problems have largely been solved by invention, engineering acumen and care in construction. They have also been exceeded; we have made it possible to build curves beyond the imagination of architects working just a few decades ago. Architecture has become sensational.

But while spectacle astonishes and entertains, and through photography grabs media attention, it is never a totally satisfying aspiration for architecture. Complete architecture does not merely entertain us, provide us with astonishing things to look at. Architecture is the art that accommodates. It embraces as well as struts. At its best it includes us as ingredients rather than consigning us to the role of adulatory spectators.

I embarked on this Notebook to explore the roles and possibilities of the curve in architecture. I began as one of those architects who temperamentally and through the conditioning of their own education and experience treat curves with suspicion. But nevertheless I wanted to 'give curves a chance'. I have enjoyed the exploration. I have come to understand that the magic and seduction, the invention and the achievement of curves has been part of architecture for a very long time. Curves can be practical but they are also poetic. Contemporary developments do not represent a radical new world but are part of a historical continuum with deep roots. It is part of our nature to want to transcend nature. It is also part of our nature to want to emulate nature. Curves play their parts in both these ventures.

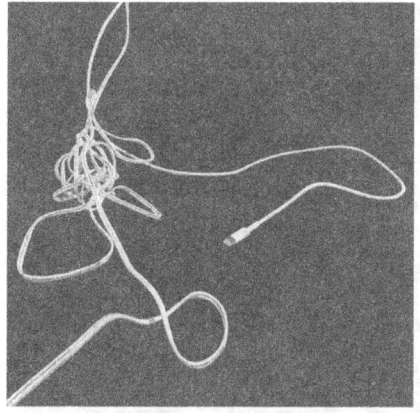 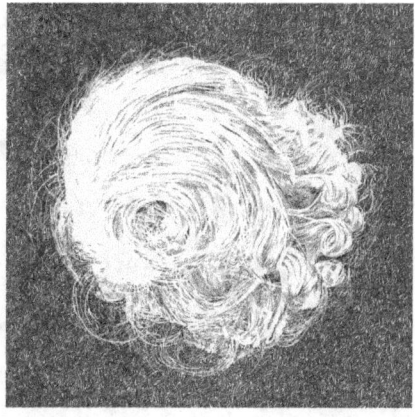

... a twisted root; rumpled fabric; tangled wires; a child's curls. Where can we go next?

It is a cliché to suggest the world would be a dull place if ruled only by reason and common sense. Presumed rules, conventions and restraints are to be tested, broken, transgressed... There may be pitfalls, but refreshment is the reward.

If I have any lingering reservations about the seduction of curves it is not because they subvert the geometry of making or seek to defy gravity. Those are heroic traits. It is because, in that general counterpoint between melody and rhythm (see pages 25–60), extravagantly curvaceous architecture can be self-regarding, reneging on its responsibility for providing a rhythm against which we can dance to our own human tunes, in favour of 'hogging the dance floor' itself. It is a frame asserting itself as the picture. We are no longer ingredients of such architecture but held at arm's length as spectators. We, the should-be players in the drama, are restricted to the role of audience, alienated by architectural frameworks that should embrace us.

Maybe this is the next challenge for curvy architecture: to find ways of including human beings as contributing partners in their dance: snug molluscs in tailor-made shells; egg-cracking chicks cupped in moss-lined nests; primates swinging from branch to curving branch swaying in the breeze; colourful fish swimming in and out of coral reefs... Ingredients again.

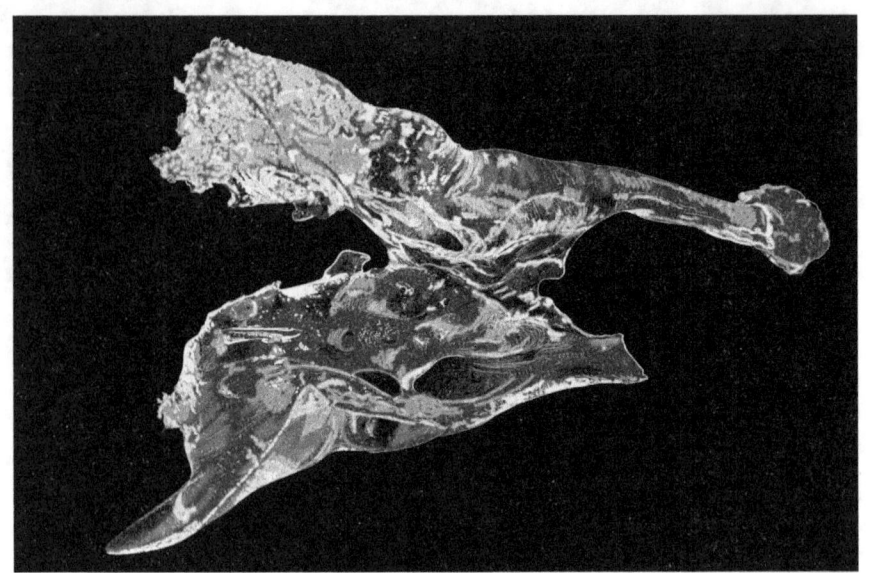

The curves of both ice and shells are beautiful. But shells are architectural because they are for inhabitation too. The touchstone of architecture is identification of place…

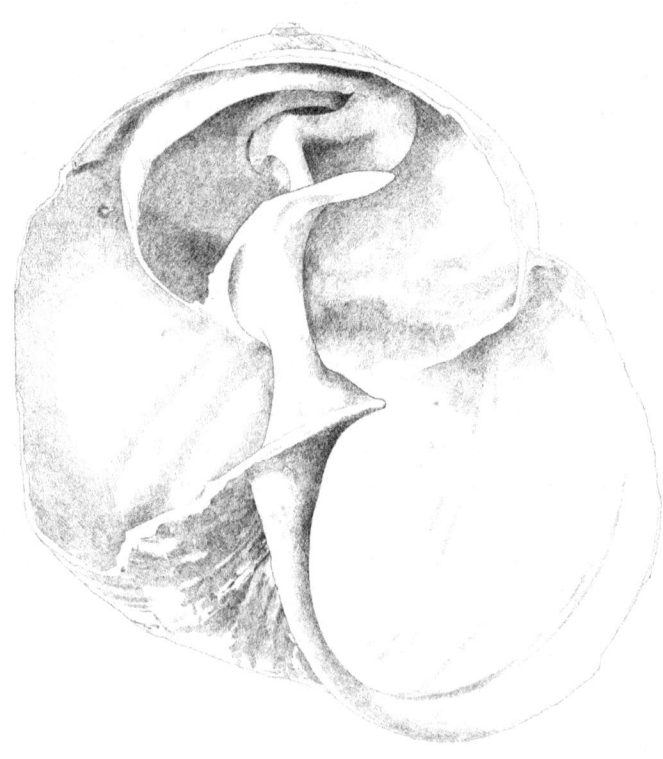

ACKNOWLEDGEMENTS

I am very grateful, for contributions and thoughts of various kinds, to: Jonathan Adams, Alice Brownfield, Professor Wayne Forster, Dr Wassim Jabi, Professor Stephen Kite, Ted Landrum, David McLees, Alan Paddison, Richard Powell, Sue Ryrie, Emily Stanley. Some of the above have prompted me to do this Notebook by their own aversion to curves in architecture!

As always I am thankful for the support and encouragement of various people at Routledge, especially Fran Ford (who, from my long list, selected the topics for the first few of these Notebooks), Jennifer Schmidt, Trudy Varcianna and Christina O'Brien.

I am grateful to the Kunstbibliothek of the Staatliche Museen zu Berlin, and Anne Schultze of their agents bpk-Bildagentur, for permission to use Erich Mendelsohn's sketch of the Einstein Tower (as photographed by Dietmar Katz) on page 45.

And as always, I am grateful to my family, some of the newer members of which will feature in another Notebook in this series – *Children as Place-Makers*.

BIBLIOGRAPHY

Chris Abel – *Manikata Church*, AD Academy Editions, London, 1995.
Robert William Billings – *The Power of Form Applied to Gothic Tracery*, William Blackwood and Sons, Edinburgh, 1851.
William Blake – *Jerusalem: The Emanation of the Giant Albion*, 1804–20.
Auguste Choisy – *Histoire de l'architecture* (1899), Éditions Vincent, Fréal, Paris, 1964.
Theodore Andrea Cook – *The Curves of Life*, Constable, London, 1914.
John Donat, editor – *World Architecture 2*, Studio Vista, London, 1965.
James Fergusson – *The Illustrated Handbook of Architecture*, John Murray, London, 1859.
James Fergusson – *History of the Modern Styles of Architecture*, John Murray, London, 1862.
E.M. Forster – *Howards End* (1910), Penguin, London, 2012.
Julien Guadet – *Éléments et théorie de l'architecture* (4 volumes), Librairie de la Construction Moderne, Paris, 1894.
Alexander Holton – 'The Working Space of the Medieval Master Mason: the Tracing Houses of York Minster and Wells Cathedral', in *Proceedings of the Second International Congress on Construction History*, Volume II, 2006, pp. 1579–97, available at: arct.cam.ac.uk/Downloads/ichs/vol-2-1579-1598-holton.pdf (March 2018)
Wassim Jabi – *Parametric Design for Architecture*, Laurence King, London, 2013.
Anthony Johnson – *Solving Stonehenge*, Thames & Hudson, London, 2008.
Maxine Hong Kingston - 'No Name Woman', in *The Woman Warrior*, Allen Lane, London, 1977.
Greg Kot – 'Space Oddities: David Bowie's Hidden Influences', 2016, available at: bbc.com/culture/story/20160108-space-oddities-david-bowies-hidden-influences (December 2018).
Robert McCarter – *Frank Lloyd Wright*, Phaidon, London, 1997.
Franz Sales Meyer – *A Handbook of Ornament* (1894), Omega Books, London, 1987.
Iorweth Peate – *The Welsh House* (1940), Brython Press, Liverpool, 1946.
Bernard Rudofsky – *Architecture Without Architects*, Academy Editions, London, 1964.
John Ruskin – *The Elements of Drawing* (1857), Dover, New York, 1971.
Vincent Scully – *The Earth, the Temple, and the Gods* (1962), Yale University Press, New Haven and London, 1979.
Abbot Suger, trans. Panofsky – *On the Abbey Church of St.-Denis and its Art Treasures* (12thC), Princeton U.P., 1946.
D'Arcy Thompson – *On Growth and Form* (1917, 1942), Cambridge U.P., 1966.
Georg Gottlob Ungewitter and K. Mohrmann – *Lehrbuch Der Gotischen Konstruktionen* (two volumes; 1858), Tauchnitz, Leipzig, fourth edition, 1901.
Piers Taylor – 'The House That 100k Built', BBC, 23 March 2017.
Simon Unwin – *Analysing Architecture* (1997), Routledge, Abingdon, fourth edition, 2014.
Simon Unwin – *Children as Place-Makers* (Analysing Architecture Notebook Series), Routledge, Abingdon, 2019.
Simon Unwin – *Metaphor* (Analysing Architecture Notebook Series), Routledge, Abingdon, 2019.
Simon Unwin – *Twenty-Five Buildings Every Architect Should Understand*, Routledge, Abingdon, 2015.
Vitruvius, trans. Morris Hickey Morgan (1914) – *The Ten Books on Architecture* (1stC BCE), Dover, New York, 1960.
Richard Weston – *Utzon*, Edition Bløndal, Hellerup, 2002.
Henry Wotton – *The Elements of Architecture* (1624), Gregg International, Farnborough, 1969.

INDEX

3d printing 92, 173
4A Four Architects 52
30 St Mary Axe, London (Foster) 118

Aalto, Alvar 135, 136, 169, 170
Abramovic, Marina 118
abstract expressionism 33, 44
Adobe Photoshop 2, 6
Adobe 'smoothing' facility 2
advertisement 39
aerodynamic curves 139, 140
aeroplane design 91, 139
'Air Beam' technology 85
AirSculpt 93
Al-Aqsa Mosque, Jerusalem 74, 152
algorithms 100
alignment 8, 18
Andreu, Paul 114
'Another Place' (Gormley) 8
anthropomorphic curves 117, 129
Apple HQ, Cupertino (Foster) 145
Aquarium, Batumi (Larsen) 126
Aquatic Centre, London Olympics (Zaha Hadid) 86
arc 73, 121
arch 61, 67, 68, 75, 171
Archdaily 20, 173
'architect of the universe' 160
Architects Co-Partnership 84
architectural promenade 54
Archway of Ctesiphon 73
arena 41
Ariston of Paros 34
artifice 172
Artist's Studio, Somerset (Lea) 90
Arup, Ove 84
ashlar 11
'ash tree of the world' 122
asymmetry 33
Aurora Borealis 136
authority 3, 4, 76, 157, 168, 172
Autodesk 3ds Max 2, 6, 106
Automobile Objective (Wright) 57
axis 18, 29, 32

Bagsværd Church, Denmark (Utzon) 134
ballet dancer 35
balloon 85
Baroque curves 165
Basilica, Paestum 120
beach 43, 46
beat 25, 33
beauty 32
Beijing Sino Sculpture Landscape Engineering Company 104
bending 88, 89, 90
'Berlin Dancer' 34
Bertelsmann Verlag, Gütersloh 48
Berthier, Julien 22
Bini, Dante 85
'Binishell' roof 85

biomimicry 172
Bioscleave House (Gins and Arakawa) 49
Biosphere, Montreal (Buckminster Fuller) 102
birds 43, 139
bivouac 63, 96
Blake, William 3, 160
boat design 91
boiler house, Brynmawr Rubber Factory 84
bones 172
bonsai 37
Bonvicini, Monica 116
Bourges Cathedral tracing floor 156
box-like room 43, 46
Bramante, Donato 163
brick 12, 98, 99
Bridge of Aspiration, Royal Ballet School, London (WilkinsonEyre) 104
British Museum, Great Court roof (Foster) 103
Brunel, Isambard Kingdom 81
Bryn Celli Ddu, Anglesey 117
Brynmawr Rubber Factory (Architects Co-Partnership with Arup) 84, 137
Buckminster Fuller, Richard 102
burial mounds 117
bürolandschaft 48
Byzantine curves 149

Calatrava, Santiago 83
calculation 172
calligraphy 33, 44
Callimachus 108
'Can-Can' (Seurat) 119
Candela, Félix 83, 101
Carpenter Center for the Visual Arts (Le Corbusier) 55
Carpenter, Rhys 121
cars 139
caryatid 115
Casa Batlló, Barcelona (Gaudí) 128
Casa Milà, Barcelona (Gaudí) 78, 125
Castel Béranger, Paris (Guimard) 124
catenary 72, 73, 75, 76, 78, 79, 81
cave 118, 130, 132, 135
chain 2, 76, 77
chain models (Gaudí) 79
Chapel Lomas de Cuernavaca, Mexico (Candela) 101
Chinese subterranean structures 133
Chippendale chair 24
Choisy, Auguste 158
chord 163
choreography 31
circle 4, 63, 67, 80, 89, 144, 145, 151, 154, 163, 171
circle template 5
Circular Pavilion, Hadrian's Villa, Tivoli 148
City of Arts and Sciences, Valencia (Calatrava) 83
cleft 116
Clifton Suspension Bridge, Bristol (Brunel) 81

clouds 134, 172
cobwebs 138, 172
Colònia Güell crypt (Gaudí) 79
colonne torse 168
Columbo (Peter Falk) 7
columned hall 47
Comme des Garçons store, New York (Future Systems) 91
common sense 175
compasses 4, 19, 76, 145, 152, 154, 157, 160, 172
computer-controlled drones (Intel) 173
computer software 1, 141, 145, 172, 173
concrete 14, 30, 92
concrete shell dome 84
conical roof structure 63, 96
conic sections 80
content curves 30
context 28
convention 23
corbel dome 66
Corinthian capital 108, 109
Corinthian column 168
counterpoint 170
cranium 70
crooked 19
Crooked House, Sopot (Zaleski) 23
cruck frame 64
Cullinan, Ted 89
curly hair 26, 175
Curve Appeal Home (Wimberly, Allison, Tong & Goo) 92
curve ball 1
curved cue 7
curve generation methods 4
curve metaphor 174

dance 24, 33, 34, 41, 47, 175
dance of life 31
decadence 19, 25
defiance 171
Devil's Bridge, Kromlau 67
Dezeen 20, 173
Didyma 121
disorder 14
distortion 22, 23
distrust 19
divine geometry 165
DNA 25
Dogon shrine to the sacred feminine, Mali 118
Doha Villa (Findlay) 127
dome 171
doorway 16
doorway axis 9
Dr Chau Chak Wing Building, Sydney (Gehry) 99

earth curvature 17
École des Beaux-Arts, Paris 166
Edinburgh Sports Dome, Malvern (Godwin) 85
egg 71, 113, 114, 175
Einstein Tower (Mendelsohn) 45
elegance 13, 30, 34
'Elegy Written in a Country Churchyard' (Gray) 33
Elements of Drawing (Ruskin) viii
Elephanta cave temples, India 130
ellipse 80, 81, 84, 146, 147, 165, 171

emotion 35, 44, 117
Endless House (Kiesler) 53, 60
England, Richard 131
entasis 120, 121, 168
Epidaurus 109
erotic curves 115, 116
Euler spiral 5
ever-present melody 25
expense 24

fairground mirror 23
Fallingwater (Wright) 56
fantasy 172
fan vault 161
Farkasréti Cemetery, Budapest (Makovecz) 129
fashion design 39
fauna 25
feathers 42
fecund curves 113
Fehn, Sverre 54, 55
Feng Shui 55
Fergusson, James 69
Findlay, Kathryn 60, 127
Finnish Pavilion, New York (Aalto) 136
fish 172
fish skull 128
flatness 30
flesh 115
flexicurve 5
floor curves 49
flora 25
flowers 145
flying buttress 75
flying saucers 140
focus 18
foil 25, 27, 29, 33, 37, 41
food processor 142
Forestry Pavilion, Lapau (Aalto) 135
formwork 14, 30
Forster, E.M. viii
Foster + Partners 103, 118
four-square 9, 20
frame 1, 14, 33, 175
freedom 31
Freemasons 19, 154
'free plan' 1, 46
French curves 5
Fri og Fro ecoVillage, Denmark (Schmidt) 96
fruit 113
furniture 15, 31
fusion 32, 170
future 45, 171
Future Systems 91, 117, 118

Gammes Dynamosphériques (Laban) 44
Gateshead Millennium Bridge (WilkinsonEyre) 82
Gateway Arch, St Louis (Saarinen) 77
Gaudí, Antoni 78, 125, 128
Gehry, Frank 21, 39, 99
generation of curves 3
geodesic dome 102
geometric constructions 100
geometry 1, 14, 27, 29, 154
geometry of making 16, 20
ger 89

German Pavilion, Montreal Expo (Otto) 138
gesture 4, 44, 45, 143, 172
Gibberd, Frederick 96
Gins, Madeline and Arakawa 49
Givone, Tom 20
glass 22
Godwin, Michael 85
Golden Section 32
Gormley, Antony 8
Gothic capitals 111
Gothic curves 74, 122, 123, 153, 156, 158, 159, 161
Gournia, Crete 117
grain 30
gravity 1, 9, 10, 11, 19, 25, 76, 81
Gray, Thomas 33
Great Mosque of Samara, Iraq 57
Green Bird (Future Systems) 118
Greenland Tower, Wuhan, China (Smith and Gill) 140
grid 17, 47
Guadet, Julién 69
guardsman 35
Guedes, Amâncio d'Alpoim Miranda 'Pancho' 116
Guggenheim Museum, Bilbao (Gehry) 21
Guggenheim Museum, New York (Wright) 56, 59
Guimard, Hector 124
guna tubes 97
gymnastics 41

'habitable woman' (Guedes) 116
Hadrian's Villa, Tivoli 148
Hagar Qim, Malta 131
Hagia Sophia, Istanbul 70, 149
hairstyling 33
Hakusasonsu Garden, Kyoto 50
Handley Page biplane 141
handwriting 4
Hansen, Oskar Nikolai 93
'happy' and 'sad' curves 119
Hassan Fathy 72, 73
Hedmark Museum, Hamar (Fehn) 55
heroism 172, 175
hesitant curve 4
Heydar Aliyev Cultural Centre (Zaha Hadid) 24
Hill Holt Wood (Lincoln School of Architecture) 96
Hippodamus 17, 28
Hong Kingston, Maxine 115
Hopkins, Michael 86
horizon 33
horizontal arch 66
Hotel Marqués de Riscal, Elciego (Gehry) 39
'House That 100k Built' viii
Howards End (Forster) viii
How to Dial a Murder (Frawley) 7
hydrodynamic curves 139
hyperbola 75, 80
hyperbolic paraboloid 83, 87, 93, 100, 101, 105
hyperbolic paraboloid potato crisp 87
Hypogeum, Malta 131

iambic pentameter 33
ice 176
ice cave 136
'Imponderabilia' (Abramovic) 118
Incurvo House, Oxfordshire (James) 99

Indian Institute of Management, Ahmedabad (Kahn) 99
infinity 53
'infinity' ramp (4A Four Architects) 52
inflation 85
innate geometry 10
instruments of curvature 3
intangible walls 132
Intel drones 173
interface 25
interplay 28, 29
Ionic volute 112, 167
Iron Age house 63
irregularity 28, 144
Isaiah 19

James, Adrian 99
'Jerusalem: The Emanation of the Giant Albion' (Blake) 3
jigsaw 15
joke 23
Jorn, Asger 133

Kahn, Louis 12, 98, 99
Kasuyo Sejima 40
Kiesler, Friedrich 53
King's College Chapel, Cambridge 69, 161
Kisho Kurokawa 142
knucklebones 127
Korsgaard, Søren 105
Krzywy Domek (Zaleski) 23

Laban, Rudolf von 44
Lake Garden sculpture, Suzhou (Beijing Sino Sculpture Landscape Engineering Company) 104
Larsen, Henning 126
lattice 89, 90
Lea, David 90
leaves 42
'Le Chahut' (Seurat) 119
Le Corbusier 46, 47, 54, 55, 119, 132
Leonardo da Vinci 3, 26, 27, 32, 34
Leonardo Glass Cube (3Deluxe) 48
Leveil, J.-A. 166, 168
level 19
life drawing 115
light 122
light drawings (Picasso) 44
Lincoln School of Architecture 96
lines of force 75
Liverpool Metropolitan Cathedral (Gibberd) 96
Los Manantiales, Mexico City (Candela) 83
'Luna Pod' (Evolution Dome) 85

Maggie's Centre, Swansea (Kisho Kurokawa) 142
magic 143, 171, 174
magic wand 144
Maison La Roche (Le Corbusier) 54
Makovecz, Imre 129
Malator (Future Systems) 117
'male' and 'female' 35
Mallet-Stevens, Robert 19
Manikata Church, Malta (England) 131
mathematics 172

matrix 46, 102, 106
mechanical devices 1, 172
Media Centre, Lord's Cricket Ground, London (Future Systems) 91
melody 1, 25, 33, 47, 170, 175
Mendelsohn, Erich 45, 106
Mensch und Kunstfigur (Schlemmer) 35
messiness 14, 16
Meyer, Franz S. 159, 160
Mezquita de Córdoba, Spain 69, 150
Mies van der Rohe 24
mihrab, Mezquita de Córdoba 150, 151
Miletus (Hippodamus) 17, 28
mirror line 18
Modernism 1
Momo, Giuseppe 52
monocoque 91
Moore, Henry 127
mother 117
Mott, Hay and Anderson 82
movement 25, 31, 34, 35, 36, 41, 44, 51
Mowry, James 101
Mozart 47
murmuration of starlings 173
muscles 120
music 47

narrative 23, 25, 33
National Centre for the Performing Arts, Beijing (Andreu) 114
Nationalgalerie, Berlin (Mies) 24
nature 1, 8, 12, 26, 29, 38, 107, 128, 172
neatness 15
Neumann, Balthasar 165
New Gournia, Egypt (Hassan Fathy) 72
Niemeyer, Oscar 140
Niterói Contemporary Art Museum (Niemeyer) 140
Nolli, Giambattista 28
'non-uniform rational B-spline' (NURBS) 6, 106
Northern Lights 136
Norwegian Wild Reindeer Centre (Snøhetta) 38
Notre-Dame d'Avy, Charente Maritime, France 152
Notre Dame, Paris 160
NURBS surface 6, 106
nurture 114

obfuscation of space 165, 171
Okurayama Apartments, Japan (Kasuyo Sejima) 40
Olympia, Greece 67
orbit 25, 144
orchestration of curves 143, 169, 170
Organic Cube, Copenhagen (Korsgaard) 105
ornamental curves 1, 165
orthogonal 7, 16, 17, 19, 20, 25, 27, 29, 30, 32, 33, 38, 46, 170
Oscar Niemeyer Museum, Curtiba (Niemeyer) 140
ostentation 24
Ostia 68
Otto, Frei 138
ovoid 114

pair of compasses 4, 19, 76, 145, 152, 154, 157, 160, 172
Palazzo Chiaramonte-Steri, Sicily 74

palimpsest 157
Palladio, Andrea 162
Pantheon, Rome 70
parabola 72, 73, 75, 80, 81, 82, 88
paradise 19, 149
parallel 13, 15
parametric software 2, 6, 20, 76
Paris Metro stations (Guimard) 124
Parthenos, Neapolis 112
pathway 42, 43, 46, 47, 48, 50, 51, 53, 54, 153, 171
Pausanias 109
Peak Cavern, Derbyshire 130
pebbles 11, 33, 126, 172
pegs and string 4, 61, 172
pelvis 129
Pergamon 68
phallic curves 118
Phrasikleia Kore (Ariston of Paros) 34
Picasso 44
picture frame 27
Pirelli Tower (Ponti) 36
Platonic geometry 29, 32, 56, 157, 161
plinth 27
plughole 142
poetic curves 174
poetry 30, 33, 37
Polish Pavilion, Izmir International Fair (Hansen) 93
Pollock, Jackson 44
Polykleitos 109
Ponti, Gio 36
pool 7
Popular Theatre, Niterói (Niemeyer) 140
pregnancy bump 113
prehistoric curves 146
prism 31
procreation 113
proportion 27
publicity 171
punchline 23

quatrefoil 158
question-mark 115

Raffles City, Hangzhou (UNStudio) 36
raindrops 42
ramp 57
random curve 3
reciprocal roof 96
rectangular 1, 7, 14, 15, 16, 20, 31, 41, 95, 171, 172
rectitude 19
Red Blue Chair, left-handed (Berthier) 22
Red Blue Chair (Rietveld) 22
reflections 125
refuge 38
regularity 28, 30, 32
renewal 113
rhythm 1, 33, 47, 175
ribbons 39
ribcage 129
Rietveld, Gerrit 22, 37
right 19, 27
right angle 7, 12, 15, 19, 20, 95
rock-cut architecture 130
rock strata 3
Rococo curves 165

roller coaster 52
Roman arches 68
Roman curves 148
Romanesque curves 152
Roman vaulting tubes 97
Ronchamp chapel (Le Corbusier) 119, 132
root 175
rope bridge 81
rose window 160
roundhouse 63, 96
Rudofsky, Bernard 88
ruler 3
Ruskin, John viii, 107

Saarinen, Eero 77, 137
Sackler Gallery, London (Zaha Hadid) 94
Sagrada Familia, Barcelona (Gaudí) 79
sail 43
Schlemmer, Oskar 35
Schmidt, Poula-Line 96
Schröder House (Rietveld) 37
scribble 53
Scully, Vincent 117
sculptural curves 170, 171
sculpture 115
sea 43
seashell 112
seduction 174
sensuous curves 115
set square 19
Seurat, Georges 119
sexual characteristics 117
Shaker chair 24
sharks 139
shed 13
shell 3, 71, 134, 137, 167, 172, 175, 176
shrine 33
Silkeborg Museum (Utzon) 133
six directions 9
skeleton 129
skin 91
skull 71, 128
sky 134
Smith, Adrian and Gill, Gordon 140
snakes 42
Snøhetta 38
snow cave 38
'Soave sia il vento' (Mozart) 47
Sodagar, Behzad 96
Solomonic column 168
space of life 40
space-time 53
spatial organisation 31
special brick 99
spectacle 174
spine 129
spiral 59, 112
spiral galaxy 142
spiral ramp, Vatican (Momo) 52
spiral stair 51
'spline' curve 6
square 13, 14, 15, 19, 27
squiggle 4
stairs 52
standing to attention 35, 44

Stansted Airport (Foster) 50
Stansted Airport (World Duty Free Group) 50
steel 14
stepping stones 51
St Fagans National Museum of History, Cardiff 63
stillness 44
St Mary, Snettisham 123
Stonehenge 145, 147
'Stonewall 3' (Bonvicini) 116
St Peter's Basilica, Rome 28
straight 1, 7, 10, 12, 13, 14, 15, 18, 19, 35, 120
straight edge 3
straight line 1, 8, 19, 20, 31, 35, 41, 95, 100, 130, 165, 171, 172
Strasbourg Cathedral 75, 157
strength 120
stretched fabric 93, 94, 141
Stretch Marquees and Fabric Structures 93
'stretch pod' (Stretch Marquees and Fabric Structures) 93
string 2, 4, 72, 144
stroll garden 50
structure 10, 15, 61, 62
submarines 139
subterranean structures 133
Suger, Abbot 122
suspension 171
suspicion 19
swim 43
swing balls 1
switchback 42
symbiosis 16
symmetry 18

Taj Mahal, India 164
Tāq Kasrā (Archway of Ctesiphon) 73, 75
Taylor, Piers viii
teepee 96
Tempietto, Rome (Bramante) 163
templates 5, 172
temple 9, 62
Temple of Apollo, Bassae 29, 109, 112
Temple of Hera (Basilica), Paestum 120
Temple of Isis, Philae 110
Temple of Mars Ultor, Rome 121
tendrils 3, 124, 172
tension wires 87
tent 88, 93
termites 92
'The Dryad's Waywardness' (Ruskin) 107
three-centred arch 72
Thymele, Epidaurus (Polykleitos) 109
Tiber 28
timber 13, 14, 30
tokonoma 33, 37
topography 25, 42
tracery 122, 123, 158, 159
tracing floor 156
traditional house, Iraq 88
Trajan's Column 51
Treasury of Atreus, Mycenae 66
tree 37, 65, 122, 123, 153, 172
trefoil 158
trick drawing 119
Truss Wall House (Ushida Findlay) 60

'truth' 10, 22
turbulence 141
Tuscan column 168
TWA Flight Center, New York (Saarinen) 137
twist 23
Twisted Farmhouse (Givone) 20
'Two Large Forms' (Moore) 127
Tyne Bridge (Mott, Hay and Anderson) 82

Ungewitter, Georg Gottlob 75, 111, 158
UNStudio 36
upright 35
Urizen (Blake) 160
Ushida Findlay 60
Utopia 19
Utzon, Jørn 133, 134

Valley Christian Reformed Church, Kattelville, NY (Mowry) 101
vase (Aalto) 135
vault 61, 69, 75, 83, 97, 153, 161, 171
vaulting tubes 97
V.C. Morris Store, San Francisco (Wright) 58
Velodrome, London Olympics (Hopkins) 86
vertical 7, 10, 34
Vierzehnheiligen, Bavaria (Neumann) 165
Vignola, Giacomo Barozzi da 121, 168
Viipuri Library, Finland (Aalto) 135
Villa Mairea, Finland (Aalto) 169, 170
Villa Noailles (Mallet-Stevens) 19
Villa Rotonda, Vicenza (Palladio) 162, 163
Villa Savoye (Le Corbusier) 54
Vinarium Tower, Lendava, Slovenia (Virag and Rajšter) 101
Virag, Oskar and Rajšter, Iztok 101
vitality 34

Vitruvian Man (Leonardo) 32
Vitruvius 32, 34, 108, 121
vortex 142, 172
Vuoksenniska Church, Imatra (Aalto) 136

Wacom computer touch screen 2
walking 42
wall 11
Ward Willits House (Wright) 56
warping 13, 19
water 22
waves 3, 140
Weald and Downland Open Air Museum, England (Cullinan Studio) 89
Wells Cathedral, England 154
Weston, Richard 133
wigwam 89
WilkinsonEyre 82, 104
Wimberly, Allison, Tong & Goo 92
wind 39, 88
Winter Olympic Games, Pyeongchang, Korea 173
Winton Gallery, London (Zaha Hadid) 141
wires 175
womb 117
Women's Baths, Hadrian's Villa, Tivoli 148
Woodhenge 146
Wotton, Henry 64
wrapping 39
Wright, Frank Lloyd 56

York Minster tracing floor 156, 157
yurt 89

Zaha Hadid 24, 86
Zaleski, Szotynscy 23
zoomorphic curves 129

'I feel confident imposing change on myself. It's a lot more fun progressing than looking back. That's why I need to throw curve balls.'

David Bowie, quoted in Greg Kot – 'Space Oddities: David Bowie's Hidden Influences', 2016.

For Product Safety Concerns and Information please contact our EU
representative GPSR@taylorandfrancis.com
Taylor & Francis Verlag GmbH, Kaufingerstraße 24, 80331 München, Germany

www.ingramcontent.com/pod-product-compliance
Lightning Source LLC
Chambersburg PA
CBHW051100230426
43667CB00013B/2381